SIGNIFICANT CHANGES TO THE
INTERNATIONAL PLUMBING CODE, INTERNATIONAL MECHANICAL CODE, AND INTERNATIONAL FUEL GAS CODE

2006 EDITION

THOMSON

∗

DELMAR LEARNING ™

Australia   Canada   Mexico   Singapore   Spain   United Kingdom   United States

THOMSON
DELMAR LEARNING

Significant Changes to the International Plumbing Code, International Mechanical Code, and International Fuel Gas Code 2006 Edition
Robert G. Konyndyk, Robert Guenther, and Stephen A. Van Note

Vice President, Technology Professional Business Unit:
Gregory L. Clayton

Product Development Manager:
Ed Francis

Editorial Assistant:
Jaclyn Ippolito

Director of Marketing:
Beth A. Lutz

Executive Marketing Manager:
Taryn Zlatin

Marketing Specialist:
Marissa Maiella

Director of Production:
Patty Stephan

Production Manager:
Andrew Crouth

Content Project Manager:
Kara A. DiCaterino

Art Director:
Robert Plante

Library of Congress Cataloging-in-Publication Data:
Card Number:

Application Submitted

ISBN: 1-4180-5382-1

NOTICE TO THE READER

Contents

Preface

Code officials, design professionals, and others involved in the building construction industry recognize the need for a modern, up-to-date building code addressing the design and installation of building systems, including plumbing, mechanical, and fuel gas systems, through requirements emphasizing performance. The 2006 editions of the *International Plumbing Code*® (IPC), *International Mechanical Code*® (IMC), and *International Fuel Gas Code*® (IFGC) are intended to meet these needs through model code regulations that safeguard public health and safety in all communities, large and small. The IPC/IMC/IFGC are kept up-to-date through the open code-development process of the International Code Council (ICC). The provisions of the 2003 editions, along with those code changes approved through 2005, make up the 2006 editions.

The ICC, publisher of the I-Codes, was established in 1994 as a nonprofit organization dedicated to developing, maintaining, and supporting a single set of comprehensive and coordinated national model building construction codes. Its mission is to provide the highest quality codes, standards, products, and services for all concerned with the safety and performance of the built environment.

The IPC, IMC, and IFGC are three of the 14 International Codes® published by the ICC. These comprehensive codes establish minimum regulations for plumbing, mechanical, and fuel gas systems by means of prescriptive and performance-related provisions and are founded on broad-based principles that make possible the use of new materials and new system designs. The IPC, IMC, and IFGC are available for adoption and use by jurisdictions internationally. Their use within a governmental jurisdiction is intended to be accomplished through adoption by reference, in accordance with proceedings establishing the jurisdiction's laws.

The purpose of *Significant Changes to the International Plumbing Code, International Mechanical Code, and International Fuel Gas Code 2006 Edition* is to familiarize plumbing and mechanical officials, building officials, fire officials, plans examiners, inspectors, design professionals, contractors, and others in the construction industry with many of the important changes in the 2006 IPC/IMC/IFGC. This publication is designed to assist those code users in identifying the specific code changes that have occurred and, more important, in understanding the reasons behind the changes. It is also a valuable resource for jurisdictions in the code-adoption process.

Only a portion of the total number of code changes to the IPC/IMC/IFGC are discussed in this book. The changes selected were identified for a number of reasons, including their frequency of application, special significance, or change in application. However, the importance of those changes not included is not to be diminished. Further information on all code changes can be found in the Code Changes Resource Collection, published by the ICC, which provides the published documentation for each successful code change contained in the 2006 IPC and 2006 IMC.

Throughout this significant changes book, each change is accompanied by a photograph, an application example, or an illustration to assist and enhance the reader's understanding of the specific change. A summary and discussion of the significance of the changes are also provided. Each code change is identified by type, be it an addition, modification, clarification, or deletion.

The code change itself is presented in a format similar to the style utilized for code-change proposals. Deleted code language is shown with a strike-through, whereas new code text is indicated by underlining. As a result, the actual 2006 code language is provided, as well as a comparison with the 2003 language, so the user can easily determine changes to the specific code text.

The plumbing code changes focus on specific details in their own context. A Reader's Note explanation is added occasionally to provide greater clarification and operational insights to the reader. These elements point out organization issues and serve as a reminder to the reader to place identified code changes together.

As with any code-change text, *Significant Changes to the International Plumbing Code, International Mechanical Code, and International Fuel Gas Code 2006 Edition* is best used as a study companion to the 2006 IPC, 2006 IMC, and 2006 IFGC. Because only a limited discussion of each change is provided, the code itself should always be referenced in order to gain a more comprehensive understanding of the code change and its application.

The commentary and opinions set forth in this text are those of the authors and do not necessarily represent the official position of the ICC. In addition, they may not represent the views of any enforcing agency, as such agencies have the sole authority to render interpretations of the code. In many cases, the explanatory material is derived from the reasoning expressed by the code-change proponent.

Acknowledgments

Robert G. Konyndyk, author of the IPC section, thanks the staff members of the ICC for their assistance in the preparation of the text.

Assembling the text with its code insights would have been impossible without permission and research help from the ICC. Bob especially thanks Hamid Naderi, ICC staff engineer, for his guidance and overall involvement throughout the writing process, including authoring several documents. His patience and expertise are greatly appreciated.

The plumbing profession, with its emphasis on providing a safe, healthy environment for citizens, has been motivated by the manufacturing industry. Manufacturers play an extremely important role in driving the code industry forward with their products. A list of those companies who have voluntarily provided graphics follows:

Airxchange, Inc.

Alsons Corp.

American Society of Sanitary Engineering (ASSE)

Aquatherm

Best Bath Systems

Burke Agency, Inc.

Cast Iron Soil Pipe Institute (CISPI)

Charlotte Pipe and Foundry Company

Crane Plumbing

Delta Faucet Co.

Genova Products

George Fischer, Inc.

Josam Co.

Kohler Co.

NSF International

Red-White Valve Corp.

Schier Products Co.

Sheet Metal and Air Conditioning Contractors' National Association (SMACNA)

Studor, Inc.

Symmons Industries, Inc.

Uponor (formerly Wirsbo)

Vanguard Piping Systems

Waterless Co. LLC

Watts

Wilkins

Zoeller Pump Co.

Bob also thanks staff members at Delmar Learning for their true development force behind this book. He extends his gratitude to Ed Francis, Product Development Editor, and Sarah Boone, Editorial Assistant, for their professional guidance. Bob believes that the behind-the-scenes book development is an important supplement to the authors' efforts.

Bob is grateful to his wife, Charmaine, for her support and encouragement to move forward in the contracting field and experience the many technical facets of the plumbing profession.

Robert Guenther, author of the IMC section, extends his gratitude and thanks to Mark Johnson and Hamid Naderi, both members of the ICC staff. Robert believes that this book would not have been written if it were not for the innovative thinking and foresight of these two individuals. He feels that without the efforts and cooperation of the people, too numerous to mention, who take the time and trouble to submit code changes and to assist the committees and ICC staff responsible for processing the changes, none of this would have been possible.

Members of the HVAC profession have been of great assistance in providing graphics and other technical assistance for this book.

Special thanks are given to Lisa Valentino for her assistance with the typing and editing of the book.

Stephen A. Van Note, author of the IFGC section, graciously thanks Doug Thornburg, of the ICC, for the opportunity to work on this project and for his guidance and trust. Stephen also thanks Hamid Naderi of the ICC for his inspiration, high standards for creating excellent materials, and tireless efforts in editing and assembling all the pieces into a coherent work. He feels as though it was a privilege to work with both Doug and Hamid.

About the Authors

Robert G. Konyndyk

Chief, Plumbing Division

Bureau of Construction Codes

Robert G. Konyndyk is Chief of the Plumbing Division within the Bureau of Construction Codes, Department of Labor and Economic Growth, State of Michigan. Mr. Konyndyk plans, organizes, directs, and controls a statewide license program that encompasses over 15,000 professionals. In addition, Mr. Konyndyk manages state field inspectors and oversees plumbing code development and administration and product acceptance. Prior to state employment, he owned and operated a plumbing contracting firm for 10 years, serving as its licensed master plumber and licensed mechanical contractor. Mr. Konyndyk started from the ground up, so to speak, when he was "buried alive" during an airport excavation his first day in a national union apprentice training program.

Mr. Konyndyk also served on active duty in the Air Force as a computer repairman after working in his father's contracting business. He has several licenses and certifications and focused on the plumbing profession with its technical merits in his formal education. He has served on numerous plumbing code committees and in associations such as the American Society for Testing and Materials International (ASTM),

ICC, and National Sanitation Foundation International (NSF), dealing with public health and plumbing issues since 1985. Mr. Konyndyk feels his greatest accomplishment was being one of three representatives chosen to initially develop the International Plumbing Code (first draft, Joint Model Plumbing Code) in 1994, as the Building Officials and Code Administrators International representative. He enjoys teaching code as a state administrator and has traveled as an instructor to several states for different code groups. His goal has been to contribute toward greater uniformity and stability in national code issues.

Robert Guenther

Senior Technical Staff

International Code Council

Robert Guenther is a Senior Technical Staff member for the ICC in Whittier, California. He provides technical assistance to the members of the ICC on the International Mechanical, Fuel Gas, and Plumbing Codes. Mr. Guenther also develops and reviews technical publications that involve mechanical, fuel gas, and plumbing material. He presents mechanical and fuel gas code seminars throughout the country. Prior to joining the ICC staff, Mr. Guenther was the Mechanical Official for Long Beach, California, where he worked for 24 years, and prior to that was a Mechanical Inspector for the City of Los Angeles for 5 years. Mr. Guenther has taught the mechanical code in community colleges in Southern California and has presented mechanical code seminars for 30 years. He has served on the code change committees for the IMC and Uniform Mechanical Code (UMC) and was on the committee that formed the original IMC.

Stephen A. Van Note

Building Official

County of Linn, Iowa

Stephen A. Van Note, CBO, is Building Official for the County of Linn, Iowa. He is responsible for the administration of the county's adopted construction codes and directs the work of three combination inspectors and two permit technicians. A certified building official, he also holds certifications in the categories of plans examiner and building, electrical, mechanical, plumbing, and combination inspector. Prior to his 15 years in code administration and enforcement with Linn County, Mr. Van Note worked for 20 years in the construction field, including project planning and management for residential, commercial, and industrial buildings. He serves on the boards of three chapters of the ICC: the Iowa Association of Building Officials, Hawkeye State Fire Safety Association, and ICC Region III Upper Great Plains Chapter.

SIGNIFICANT CHANGES TO THE

INTERNATIONAL PLUMBING CODE, INTERNATIONAL MECHANICAL CODE, AND INTERNATIONAL FUEL GAS CODE

2006 EDITION

PART 1

International Plumbing Code

Chapter 1 Through Appendix G

Chapter 1 of the *International Plumbing Code* ® clarifies how the code will be enforced by the code official. Definitions of plumbing code terminology are found in Chapter 2. General regulations in Chapter 3 identify requirements not listed in other code chapters, such as testing and inspections. Fixtures and water heaters are addressed in Chapters 4 and 5, respectively. Chapters 6 and 7 regulate water and drainage piping systems in that order. Indirect/Special Waste is referenced in Chapter 8. Chapter 9 details acceptable venting methodologies with in-depth piping arrangements. The provisions for traps with various receptors are found in Chapter 10. Storm drainage, with its piping collection system, is covered by Chapter 11. Installation and standards are identified with clear guidelines in Chapter 12 (medical gas piping) and Chapter 13 (all standard references). Appendices A through G cover nonmandatory provisions for permit fees, rainfall rates, new gray water systems, degree design temperatures, a water sizing method, and structural protection methodology. ■

202
Approved (Definitions)

202
Branch Interval (Definitions)

202
Flow Control (Definitions)

202
Grease Interceptor (Definitions)

202
Approved (Definitions)

CHANGE TYPE. Clarification

CHANGE SUMMARY. A definition has been revised to clarify the code use of the term *approved.*

2006 CODE: Approved. ~~Approved by~~ <u>Acceptable to</u> the code official or other authority having jurisdiction.

CHANGE SIGNIFICANCE. The modification further clarifies the intent of the definition of approved to be consistent with the definition found in the *International Building Code.* Editorially, the term being defined should not be used in the definition. The majority of the code provides installation guidance and standards conformance criteria. When this term is used it allows the enforcing authority to accept a specific installation or component as complying with the code. It identifies where the ultimate authority rests.

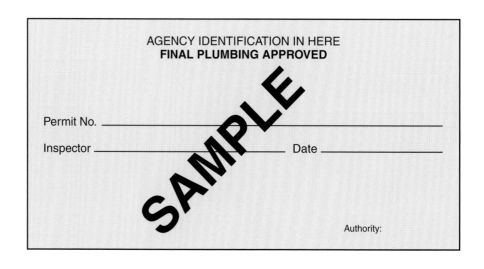

202
Branch Interval (Definitions)

CHANGE TYPE. Clarification

CHANGE SUMMARY. A definition has been revised to further clarify the distances addressed by the definition of a branch interval.

2006 CODE: Branch Interval. ~~A distance along a soil or waste stack corresponding in general to a story height, but not less than 8 feet (2438 mm), within which the horizontal branches from one floor or story of a structure are connected to the stack.~~ A vertical measurement of distance, 8 feet (2438 mm) or more in developed length, between the connections of horizontal branches to a drainage stack. Measurements are taken down the stack from the highest horizontal branch connection.

CHANGE SIGNIFICANCE. Identifying a branch interval on drainage stacks is critical to protect against overloading and increased venting problems associated with turbulent flow from branch discharges.

202 continues

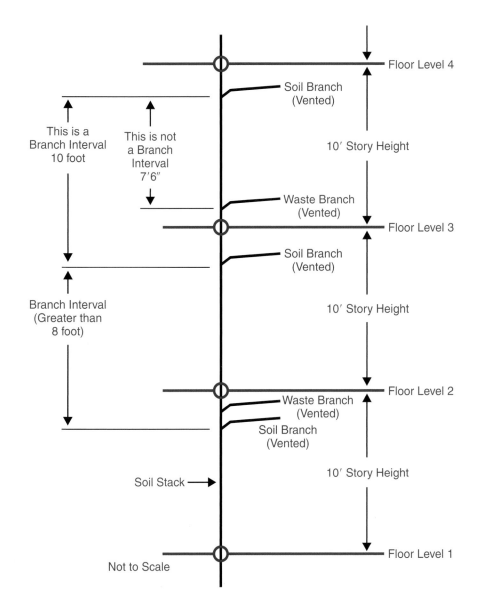

202 continued   The original text was confusing in its attempt to identify the area of a stack which received the discharge of fixtures. This revision clarifies that the stack consideration shall be based upon fixtures connecting to the stack. Further, the reference for establishing and considering branch interval numbers begins at the top of the stack, from the highest branch downward.

The size of the stacks in Table 710.1(2) is based on the number of drainage fixture units discharging into the stack at different heights, limited by the number of branch intervals. Branch intervals are also of critical importance in Section 711, offsets in drainage piping in buildings of five stories or more, and Section 914, relief vents—stacks of more than 10 branch intervals.

The plumbing industry over the years has accepted that areas in a stack 8 feet or more in developed length, between floors not having branch connections, will not be considered in Section 711 limitations.

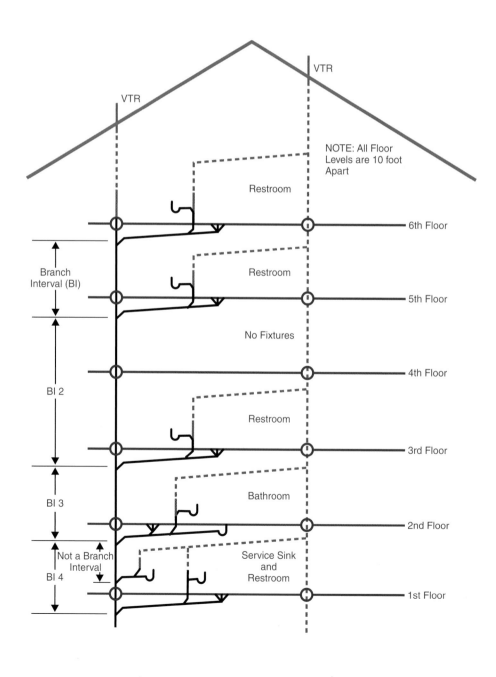

202

Flow Control (Definitions)

CHANGE TYPE.   Addition

CHANGE SUMMARY.   A definition has been added to define the device that is used on grease interceptors, formerly called grease traps.

2006 CODE: **Flow Control (Vented).**   A device installed upstream from the interceptor having an orifice that controls the rate of flow through the interceptor and an air intake (vent) downstream from the orifice which allows air to be drawn into the flow stream.

CHANGE SIGNIFICANCE.   Vented flow control devices have been provided by interceptor manufacturers for many years and are now defined in the code. The new definition now clarifies the flow control device referenced and explained in Section 1003.3.4.2. The device is also referenced in standard ASME A112.14.3, Grease Interceptors. Without the device, the interceptor would not function to its designed potential for grease retention. Grease retained in the drainage system is harmful to the plumbing and promotes drain obstructions.

The device acts as a restrictor to slow the flow, which contains grease, and allow the grease to rise in the trap holding compartment for retention. Without this restriction the retention time would be too short and the discharge velocity would be too high to separate grease from the fixture discharge. The device will have an air intake or vent and a fixed size for the particular interceptor. Using the manufacturer's installation instructions is critical for the proper operation of an interceptor.

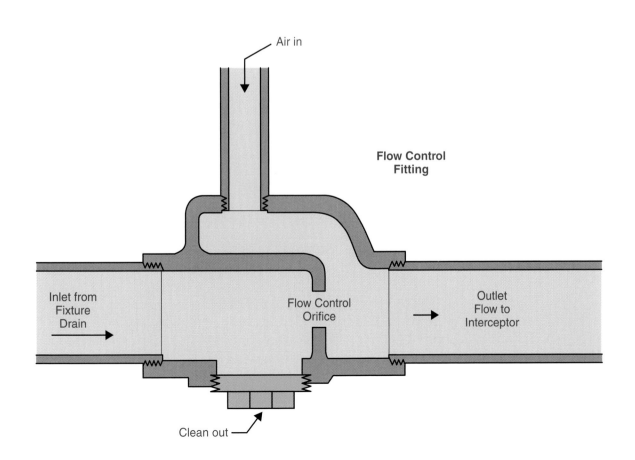

202

Grease Interceptor (Definitions)

CHANGE TYPE.   Modification

CHANGE SUMMARY.   A definition has been changed to clarify that grease traps should be called interceptors. The older term, *grease traps,* is now deleted from the definitions section of the code.

2006 CODE: Grease Interceptor.   ~~A passive interceptor whose rated flow exceeds 50 gpm (189 L/m).~~ <u>A plumbing appurtenance that is installed in a sanitary drainage system to intercept oily and greasy wastes from a wastewater discharge. Such device has the ability to intercept free-floating fats and oils.</u>

~~Grease Trap.~~   ~~A passive interceptor whose rated flow is 50 gpm (189 L/m) or less.~~

CHANGE SIGNIFICANCE.   The 2003 definition conflicts with code referenced standards ASME A112.14.3 and A112.14.4. The standards do not use the term *grease trap* and indicate all devices are considered grease interceptors. The previous reference to gpm discharges was established to differentiate between inside grease traps and outside grease interceptors.

Interceptors serving kitchen fixtures are designed to separate and accumulate grease. Grease left in the drainage system would clog the system and create performance and health concerns.

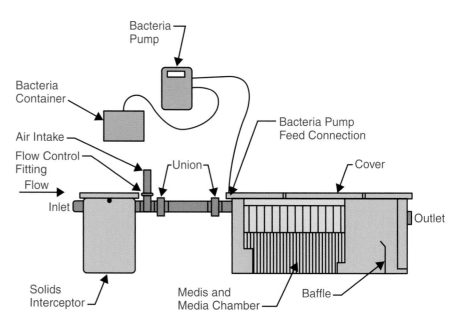

Bioremediation Grease Interceptor

CHANGE TYPE.   Addition

CHANGE SUMMARY.   The term *grease removal device* used in the code is now clarified.

2006 CODE: **Grease Removal Device, Automatic (GRD).** A plumbing appurtenance that is installed in the sanitary drainage system to intercept free-floating fats, oils, and grease from wastewater discharge. Such a device operates on a time- or event-controlled basis and has the ability to remove free-floating fats, oils, and grease automatically without intervention from the user except for maintenance.

CHANGE SIGNIFICANCE.   Automatic grease removal devices have been previously recognized by the code, and their definition is now clarified in the code. The devices shall conform to standard ASME A112.14.4, Grease Removal Devices, that is referenced in Section 1003.3.4, addressing interceptors and automatic removal devices.

202
Grease Removal Device, Automatic (GRD) (Definitions)

Automatic Grease Removal Device Control Panel

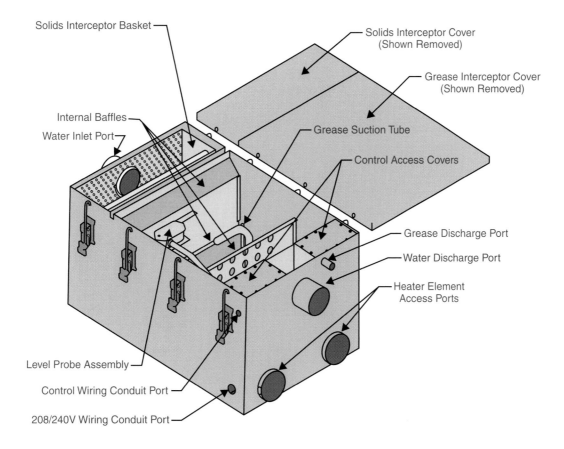

308.9

Stacks (Supports at the Base)

CHANGE TYPE.   Deletion

CHANGE SUMMARY.   The piping support information at the base of stacks has been deleted as unnecessary language.

2006 CODE: **308.9 Stacks.**   ~~Bases of stacks shall be supported by concrete, brick laid in cement mortar or metal brackets attached to the building or by other approved methods.~~

CHANGE SIGNIFICANCE.   A stack is a general term for vertical drainage, waste piping, or vent piping. The section information about requirements for the support of a pipe stack in the plumbing systems is now deleted. Generally accepted arrangements for support are clarified as requirements in sections 308.2, Piping seismic supports; 308.3, Materials; 308.4, Structural attachment; and 308.5, Interval of support.

The code for several years now has ensured the weight of the pipe and its contents are supported to avoid damage to the system. Hangers installed at each floor level, mid-floor, and branch runs distribute stack loads evenly. Hanger spacing is addressed in Table 308.5.

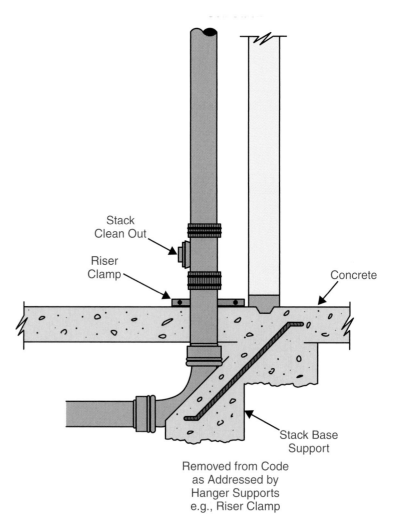

Stack Clean Out

Riser Clamp

Concrete

Stack Base Support

Removed from Code as Addressed by Hanger Supports e.g., Riser Clamp

Stacks

310.5
Urinal Partitions (Privacy)

CHANGE TYPE.　Addition

CHANGE SUMMARY.　A new section has been added to ensure privacy for occupants at a urinal.

2006 CODE:　**310.5 Urinal Partitions.**　Each urinal utilized by the public or employees shall occupy a separate area with walls or partitions to provide privacy. The construction of such walls or partitions shall incorporate waterproof, smooth, readily cleanable and nonabsorbent finish surfaces. The walls or partitions shall begin at a height not more than 12 inches (305 mm) from and extend not less than 60 inches (1524 mm) above the finished floor surface. The walls or partitions shall extend from the wall surface at each side of the urinal a minimum of 18 inches (457 mm) or to a point not less than 6 inches (152 mm) beyond the outermost front lip of the urinal measured from the finished back wall surface, whichever is greater.

> **Exception:** Urinal partitions shall not be required in a single occupant or unisex toilet rooms with a lockable door.

Reader's Note: *Another change to this section has been made by the code change process and is explained in the following significant change item.*

CHANGE SIGNIFICANCE.　The new section requires a privacy partition for occupants using a urinal. The section also identifies dimensions of the partition.

A similar matter of privacy is presently addressed for water closets in section 310.4. During the code change process it was explained that in addition to privacy, this is a sanitation and water conservation is-

310.5 continues

310.5 continued   sue. It is common where multiple urinals are located without privacy partitions that only one urinal will be used when multiple occupants enter at the same time. Occupants would elect to use the water closet, which has greater privacy, rather than using the urinal, resulting in more water usage.

Differences of opinion were expressed by several opponents of the code change, stating that urinals manufactured with small sides built into the fixture could be considered adequate. Other proponents felt those fixtures did not provide sufficient privacy. Studies have shown that a lack of privacy inhibits a person's physical ability to use restroom fixtures. The term *bashful kidneys* originated from the concept of not having adequate privacy for urinal users.

A benefit of the code change was summarized as "the codes are designed so that males and females are entitled to, and provided, the same level of amenities; this includes privacy."

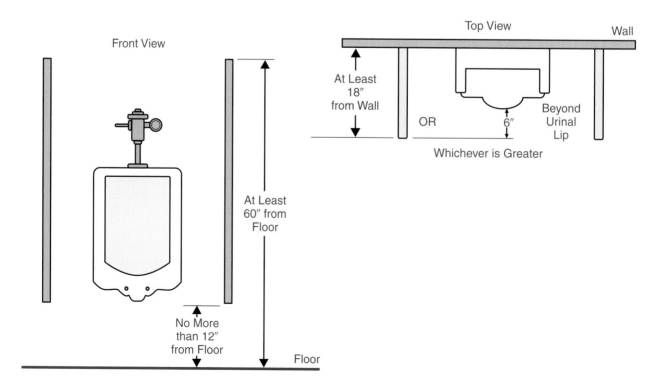

CHANGE TYPE.   Clarification

CHANGE SUMMARY.   The new second exception ensures that caregivers at child care facilities can monitor and assist small children.

2006 CODE: <u>**310.5 Urinal** ~~Privacy~~ **Partitions.**</u>   <u>Each urinal utilized by the public or employees shall occupy a separate area with walls or partitions to provide privacy. The construction of such walls or partitions shall incorporate waterproof, smooth, readily cleanable and nonabsorbent finish surfaces. The walls or partitions shall begin at a height not more than 12 inches (305 mm) from and extend not less than 60 inches (1524 mm) above the finished floor surface. The walls or partitions shall extend from the wall surface at each side of the urinal a minimum of 18 inches (457 mm) or to a point not less than 6 inches (152 mm) beyond the outermost front lip of the urinal measured from the finished back wall surface, whichever is greater.</u>

Exception:
1. <u>Urinal partitions shall not be required in a single occupant or unisex toilet rooms with a lockable door.</u>
2. <u>Toilet rooms located in day care and child care facilities and containing two or more urinals shall be permitted to have one urinal without partitions.</u>

*Reader's Note: *This significant change item explained here was added to the previous item to greater clarify the code change for the reader. The entire section above with its item #2 addition is the text found in the 2006 code document.*

310.5 continues

310.5
Urinal Partitions (Partition Day Care Exception)

Exception Number 1 Illustration

310.5 continued

CHANGE SIGNIFICANCE. The new second exception ensures that caregivers at child care facilities can monitor small children for assistance and safety purposes. Partitions might obstruct the view of the caregivers.

This code change will parallel the monitoring/safety requirements of section 310.4, the Water Closet Compartments Exception. Uniformity between code sections has always been of great importance to the code community for enforcement purposes.

312.5
Water Supply System Test (Duration of Test)

CHANGE TYPE.   Clarification

CHANGE SUMMARY.   The existing section has been changed to further expand how water supply systems are tested.

2006 CODE: 312.5 Water Supply System Test.   Upon completion of a section of or the entire water supply system, the system, or portion completed shall be tested and proved tight under a water pressure not less than the working pressure of the system; or, for piping systems other than plastic, by an air test of not less than 50 psi (344 kPa). <u>This pressure shall be held for at least 15 minutes</u>. The water utilized for tests shall be obtained from a potable source of supply. The required tests shall be performed in accordance with this section and Section 107.

CHANGE SIGNIFICANCE.   The code modification clarifies the water or air for the test pressure must be held <u>in the system</u> (emphasis added by author) for a period of 15 minutes. The previous language was often misunderstood by installers to mean that the entire test would last 15 minutes rather than the pressure being maintained for 15 minutes.

This change also provides language consistency within the testing section, 312. All other tests specified in Section 312 (such as Drainage and vent water test, Section 312.2; Drainage and vent air test, Section 312.3; Drainage and vent final test, Section 312.4; Gravity sewer test, Section 312.6; and Forced sewer test, Section 312.7) each require a 15-minute duration.

This is sufficient time to determine that a leak is not present in the system undergoing testing.

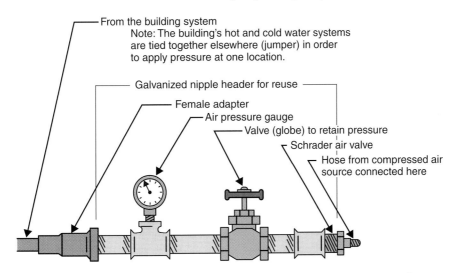

Building Water Supply System Test
with air (non-plastic system)

From the building system
Note: The building's hot and cold water systems
are tied together elsewhere (jumper) in order
to apply pressure at one location.

Galvanized nipple header for reuse

Female adapter
Air pressure gauge
Valve (globe) to retain pressure
Schrader air valve
Hose from compressed air
source connected here

Pressure not less than working pressure of system and held for
at least 15 minutes at test observation, often held overnight.

Leaks checked with soapy water solution
applied to joint connections and then checking
for air bubbles, which show leak location.

Table 403.1

Minimum Number of Required Plumbing Fixtures (Institutional Classification)

CHANGE TYPE. Modification

CHANGE SUMMARY. The Minimum Number of Required Plumbing Fixtures table for the Institutional Classification has been changed. Use Group I-4, Adult Day care and Child care, no longer includes a footnote reference to separate employee restrooms and has eliminated the requirement of a bathing facility.

2006 CODE:

TABLE 403.1 Minimum Number of Plumbing Fixtures[a]

No.	Classification	~~Use Group~~ Occupancy	Description	Water Closets (Urinals, see Section 419.2) Male	Female	Lavatories Male	Female	Bathtubs/ Showers	Drinking Fountain (See Section 410.1)	Other
5	Institutional	I-4	Adult day care and child care ~~b~~	1 per 15		1 per 15		~~1 per 15~~ ~~d~~	1 per 100	1 service sink

~~d. For day nurseries, a maximum of one bathtub shall be required.~~

*Reader's Note: *Item 1. Other changes to Table 403.1 have been made by the code change process and are explained in the following significant change items. **Item 2.** An erratum in the 2006 International Plumbing Code (first printing, dated January 2006) incorrectly deleted the "1 per 15" information for I-4, Institutional Lavatories. This information must remain as code text. Errata, which may occur, will be identified by the International Code Council on its Web site.*

EXAMPLE:

A new child care facility constructed to provide day care for children under the age of 4 has an occupancy load of 48 children. The designer has separated the employee restrooms from the children restrooms at the request of the owner/operator and will utilize various height fixtures in the children's restrooms. The owner has also requested that the children's restroom fixtures will be separated by sex due to security and potty training issues. The operators have published documents with the health department having jurisdiction that the number of care givers will be at least 12 adults at all times. These occupancy numbers are consistent with the *International Building Code* occupancy numbers based upon the designed building size and means of egress.

What are the minimum number of fixtures required for the children and what other considerations should be addressed?

SOLUTION:

Refer the Table 403.1, Minimum Number of Required Plumbing Fixtures, Institutional Classification, Occupancy I-4, which provides child care fixture numbers information.

Number of water closets: 1 water closet per 15

$$48 \text{ children} \times \frac{1}{15} = \frac{48}{15} = 3\frac{1}{5} = 4$$

Boys–2 water closets and Girls–2 water closets

Note: Three variables are often considered when computing fixture numbers when the calculations do not result in whole, even numbers, that is

1. When applying ratios listed in the table, any fraction of the quantity requires an additional fixture. OR
2. Because the fixtures are equally distributed between the sexes, an additonal fixture is required. OR
3. When a fraction of a number occurs the number is rounded up. While explanations for these three hints may vary for assembly or mixed-use groups as explained by other training aids such as the *International Plumbing Code Commentary* the outcome will remain the same.

Number of lavatories: 1 lavatory per 15

$$48 \text{ children} \times \frac{1}{15} = \frac{48}{15} = 3\frac{1}{5} = 4$$

Boys–2 lavatories and Girls–2 lavatories

Note: The variables for lavatories as expressed above for water closets apply here.

Other children's fixture considerations: Bath tubs or showers are no longer required due to the code change and it is highly unlike that a urinal will be selected as an alternative to one of the boys water closets. Further Section 310.4, Water closet compartments, Exception 2 allows one of the water closets to not have an enclosed privacy compartment. This consideration was developed for security reasons.

Building fixture considerations: A service sink and one drinking fountain are required. Employee restroom considerations: A minimum of 1 water closet and 1 lavatory are required. While the employee restrooms will be separate it is likely the owner would formally request a variance from the equal distribution of fixtures for each sex. Section 403.3, Number of occupants of each sex allows the code official to recognize all employees of the same sex based upon documentation from the owner or their agent.

Day care facilities are often the most difficult use groups to address for fixture requirements. However the communication/cooperation between the code official and plan reviewer; designer; and owner based upon practical sanitation considerations will overcome those obstacles.

Minimum Number of Required Plumbing Fixtures (Institutional classification)

CHANGE SIGNIFICANCE. The table footnote *b* required employee restrooms separate from patients' restrooms. The footnote was established for hospitals and prisons, which need separation for security or sanitation issues, not for the I-4 use group.

The information located in footnote *d* was deleted from the Table 403.1. Adult day care facilities do not bathe the adults under their care, as do facilities for overnight care. Further, footnote *d* was also deleted for child care day nurseries. The building code classifies childcare as that for children younger than 2½ years and occurring for less than 24 hours. Infant child care facilities do not bathe children in their care, and the Public Health Department requires that they have changing tables for the purpose of cleaning up. Voting consensus indicated that this was a design issue and should be left up to the owner of such facilities.

Table 403.1

Minimum Number of Required Plumbing Fixtures (Assembly Classification)

CHANGE TYPE.   Modification

CHANGE SUMMARY.   The Minimum Number of Required Plumbing Fixtures table for the Assembly Classification has been changed. Use Group A-1, Theaters, and A-3, Places of Worship, are revised to delete references to fixed seating in theaters and churches without assembly halls.

2006 CODE:

**TABLE 403.1   Minimum Number of Required Plumbing Fixtures[a]
(See Sections 403.2 and 403.3)**

Occupancy	Description
A-1	Theaters ~~usually with fixed seats~~ and other buildings for the performing arts and motion pictures
	Auditoriums without permanent seating, art galleries, exhibition halls, museums, lecture halls, libraries, arcades, and gymnasiums
A-3	Passenger terminals and transportation facilities
	Places of worship and other religious services. ~~Churches without assembly halls.~~

*Reader's Note: *Item 1. This significant change item explained here was added to the previous item and those following to greater clarify the code change for the reader.* **Item 2.** *The fixture table numbers are not shown above and have not changed. Please refer to the entire table in the code for the example solution.*

EXAMPLE:

A place of worship will be constructed which has an occupancy load of 1000. The occupancy load was established by the *International Building Code.*

What are the minimum number of fixtures required.

SOLUTION:

Refer the Table 403.1, Minimum Number of Required Plumbing Fixtures, Assembly Classification, Occupancy A-3, places of worship, formerly described as churches.

When considering assembly occupancies the total occupant load is first divided into a 50:50 ratio required by Section 403.3 of the code which states the occupancy load shall be composed of 50% of each sex.

Number of male and female occupants: 1000 occupants $\times$ 50% = 500 female occupants
500 male occupants

Number of water closets: 1 water closet per 75 female occupants

$$500 \text{ females} \times \frac{1}{75} = \frac{500}{75} = 6.66 \text{ or}$$

females = 7 water closets

1 water closet per 150 male occupants

$$500 \text{ males} \times \frac{1}{150} = \frac{500}{150} = 3.33 \text{ or}$$

males = 4 water closets

Note: When computing fixture numbers by any remainder (fraction) of the quantity requires an additional fixture.

The designer may substitute up to 67 percent of the water closets for urinals in accordance with section 419.1 of the code. While the author has placed this information under male water closets the reader should note that for many years female urinals have been available and in many cases have been accepted. Also some cultures do not accept the use of urinals due to what may be perceived as splashing problems.

Number of lavatories: 1 lavatory per 200 occupants
Number of male and female occupants: 1000 occupants $\times$ 50% = 500 female occupants
500 male occupant

1 lavatory per 200 occupants

$$500 \text{ females} \times \frac{1}{200} = \frac{500}{200} = 2.5 \text{ or}$$

females = 3 lavatories

$$500 \text{ males} \times \frac{1}{200} = \frac{500}{200} = 2.5 \text{ or}$$

males = 3 lavatories

Note: When computing fixture numbers by any remainder (fraction) of the quantity requires an additional fixture, which is similar to the water closets referenced above.

Building fixture considerations: A service sink and one drinking fountain will be required. The restrooms may be designed with floor drains for overflow conditions and would be classified as emergency floor drains. These floor drains would required trap seal primers in accordance with section 1002.4 of the code.

Minimum Number of Required Plumbing Fixtures, Assembly

CHANGE SIGNIFICANCE. The International Building Code establishes occupancy numbers and provides guidance before fixture numbers are considered. The table for A-1 occupancies, theaters, appeared unclear for theaters without fixed seating. Deleting the phrase "usually with fixed seats" will provide greater clarity.

A-3 occupancies, places of worship, are also addressed by the International Building Code occupancy information rather than consideration of assembly hall area. The identification "places of worship and other religious services" is sufficient in description in addressing the occupancy. The term *churches* was deleted for clarification.

Table 403.1

Minimum Number of Required Plumbing Fixtures (Business Classification)

CHANGE TYPE. Clarification

CHANGE SUMMARY. The Minimum Number of Required Plumbing Fixtures table for the Business Classification has been changed. Use Group B lavatory requirements have been revised to provide greater clarity in computing the number of required lavatories.

2006 CODE:

TABLE 403.1 Minimum Number of Plumbing Fixtures[a]

No.	Classification	~~Use Group~~ Occupancy	Descriptions	Water Closets (Urinals, see Section 419.2) Male Female	Lavatories Male Female	Bathtubs/ Showers	Drinking Fountain (See Section 410.1)	Other
2	Business (see Sections 403.2, 403.4 and 403.4.1)	B	Building for the transaction of business, professional services, other services involving merchandise, office buildings, banks, light industrial, and similar uses	1 per 25 for the first 50 and 1 per 50 for the remainder exceeding 50	1 per 40 for the first ~~50~~ 80 and 1 per 80 for the remainder exceeding ~~50~~ 80	(see Section 411)	1 per 100	1 service sink

***Reader's Note:** This significant change item explained here was added to the previous items and those following to greater clarify the code change for the reader.*

EXAMPLE:

A building classified as a business use group is in the design stage with an occupancy load of 1480. The office building has four floors where the owner has identified that the offices and meeting rooms will be equally distributed between the four floors.

What are the minimum number of fixtures required and how are they required to be distributed at the different floor levels?

SOLUTION:

Refer the Table 403.1, Minimum Number of Required Plumbing Fixtures, Business Occupancy.

When considering occupancies other than assembly use groups the total number of fixtures are established and then distributed between the sexes on a 50:50 ratio. Section 403.3 of the code states the occupancy load shall be composed of 50% of each sex and it is generally assumed that most businesses are composed of an equal number of each sex.

Number of water closets: 1 water closet per 25 for the first 50 and 1 per 50 for the remainder exceeding 50.

1480 occupants: $\frac{50}{25} = 2$ (for the first 50 occupants)

$1480 - 50 = 1430$

$1430 \times \frac{1}{50} = \frac{1430}{50} = 28.6$ or 29

$29 + 2 = 31$ water closets

$31 \times 50\% = 15.5$

Reader's note: Because the fixtures must be equally distributed between the sexes, an additional fixture is required.

<u>Males 16 water closets</u>

<u>Females 16 water closets</u>

Number of lavatories: 1 lavatory for the first 80 and 1 per 80 for the remainder exceeding 80.

1480 occupants $\frac{80}{40} = 2$ (for the first 80 occupants)

$1480 - 80 = 1400$

$1400 \times \frac{1}{80} = \frac{1400}{80} = 17.5$ or 18

$18 + 2 = 20$ lavatories

$20 \times 50\% = 10$

<u>Males 10 lavatories</u>

<u>Females 10 lavatories</u>

Reader's Note: The purpose the 1 water closet per 25 for the first 50 and 1 per 50 for the remainder exceeding 50 is to insure that smaller occupancies (50 or less) will have at least two water closets when a 1 to 50 would have only required 1 for this example. The same assurance is provided for lavatories such as 1 per 40 and 1 per 80.

Building fixture location considerations: The plumbing code does not have a great deal of limitations for fixture locations other than travel distance, not more than one story above and below and non-public areas.

Locations are most often based upon convenience and economic considerations. The building owner and designer have several options available to them. The code in section 403.4.1 and 403.4.2 would allow an occupant to travel up or down one flight of stairs to obtain a restroom. A men and women's restroom may be back to back on the same floor; staggered, both sexes at every other floor level; or men's, women's, men's, women's or vice versa.

The economic considerations would be based upon where the waste and water lines could be located within the 4 floor structure. Stacking the restrooms above each other may eliminate the need for enlarged walls or chases to accommodate the piping. Real estate professionals may have the lead here when they come to the design table with the term location, location, location.

Building fixture considerations: The building shall be required to have one service sink and 15 drinking fountains. Drinking fountains consideration may have alternatives from section 410.1 such as water coolers and bottled water dispensers. For this building up to 50% of the fountains, in this case 7 alternatives to fountains.

Minimum Number of Required Plumbing Fixtures, Business

CHANGE SIGNIFICANCE.   The table was revised to improve calculations in establishing lavatory number requirements for business use groups with occupant numbers between 50 and 80. Previous calculations were unclear because of the gap between 50 and 80 in the table. Designers and code officials have often questioned the former text in the table, viewing it as a typographical error.

Table 403.1

Minimum Number of Required Plumbing Fixtures (Residential Classification)

CHANGE TYPE. Clarification

CHANGE SUMMARY. The Minimum Number of Required Plumbing Fixtures table for the Residential Classification has been changed. Use Group R-2, Apartment House, and R-3, One- and Two-Family Dwellings, was revised to delete footnote e.

2006 CODE:

TABLE 403.1 Minimum Number of Required Plumbing Fixtures[a] (See Sections 403.2 and 403.3)

Occupancy	Description	Other
R-2	Apartment house	1 kitchen sink per dwelling unit; 1 automatic clothes washer connection per 20 dwelling units [e]
R-3	One- and two-family dwellings	1 kitchen sink per dwelling unit; 1 automatic clothes washer connector per dwelling unit [e]

~~e.For attached one- and two-family dwellings, one automatic clothes washer connection shall be required per 20 dwelling units.~~

***Reader's Note:** Item 1. Portions of table information for water closets, lavatories, and bathtub/showers do not change. **Item 2.** This significant change item explained here was added to the previous items and the following to greater clarify the code change for the reader.*

CHANGE SIGNIFICANCE. Footnote e is applied only to the entries R-2, Apartment House, and R-3, One- and Two-Family Dwellings, in the "other" column. The same footnote terminology is located as text in the column. The footnote and its e reference, which duplicated the requirements, have been deleted.

EXAMPLE:

What are the minimum numbers of automatic clothes washer connections in a 48-family apartment building?

ANSWER:

Refer the Table 403.1, Minimum Number of Required Plumbing Fixtures, Residential Classification, Occupancy R-2, table heading OTHER.

The table requires 1 automatic clothes washer connection to be installed for every 20 units.

$$48 \times \frac{1}{20} = \frac{48}{20} = 2.4 = 3 \text{ connections}$$

Reader's note: Code users are reminded that a portion of a fixture number representation is always "rounded up". Further remember when making calculations the preferred method is to use ratios for accuracy although most installers prefer simple division calculations.

The automatic washer connections may spread out over the building or be located in a central core washing area. Additionally, the building dwelling units may be provided with laundry facilities in each unit or some of the units, which of course would affect the mathematical outcome requirements.

Minimum Number of Required Plumbing Fixtures, Residential

CHANGE TYPE.   Modification

CHANGE SUMMARY.   The Minimum Number of Required Plumbing Fixtures table for the Assembly Classification has been changed. Use Group A-1, Theaters; A-2, Nightclubs; and A-3, Auditoriums, now have a footnote that includes outdoor seating occupants when determining the number of required plumbing fixtures.

2006 CODE:

Table 403.1

Minimum Number of Required Plumbing Fixtures, Assembly Classification (Outdoor Seating Occupants)

TABLE 403.1   **Minimum Number of Required Plumbing Fixtures[a]**
(see Sections 403.2 and 403.3)

No.	Classification	~~Use Group~~ Occupancy
1	Assembly (see Sections 403.2, 403.4 and 403.4.1)	A-1[d] A-2[d] A-3[d]

d. The occupant load for seasonal outdoor seating and entertaining areas shall be included when determining the minimum number of facilities required.

*Reader's Note: *Item 1. Portions of table information for water closets, lavatories, bathtub/showers, drinking fountain, and others do not change.* **Item 2.** *This significant change item explained here was added to the previous items to greater clarify the code change for the reader.*

CHANGE SIGNIFICANCE.   The code was unclear for A-1, A-2, and A-3 use group requirements for facilities having seasonal outdoor seating and entertaining areas. The requirement in the footnote will clarify that outdoor restaurant and bar areas must have a sufficient number of plumbing fixtures for all of the patrons.

Table 403.1 continues

EXAMPLE:

A lakefront restaurant in a northern state having an occupancy load of 138 customers and employees has been designed and provided to the code officials plan review division. The submitter informed the plan reviewer she has placed all the required fixtures in the central core male and female restrooms. Further she has added additional employee male and female restroom fixtures in the back of the kitchen area. Those fixtures are above the required fixture numbers for 138 total occupants.

The plan reviewer noted submission document contains the necessary occupancy load calculations. He further observed she had included the outdoor seating load calculations for 30 guests on the outdoor veranda in accordance with the *International Plumbing Code*, 2006 edition.

What were the minimum number of required water closets and lavatories in the central core male and female restrooms?

Table 403.1 continued

SOLUTION:

Refer the Table 403.1, Minimum Number of Required Plumbing Fixtures, Assembly Classification, Occupancy A-2, and Restaurant Description.

Occupancy load for this facility will consider the maximum number of occupants. For example the 30 outside guests will be included with the 138 indoor occupants in spite of the fact that northernmost climates may not use the outdoor seating area for a major part of the year. Remember commonly when conducting calculations for assembly occupancies the total occupant load is first separated into a 50:50 ratio. This separation is required by Section 403.3 of the code, which states the occupancy load shall be composed of 50% of each sex and then the fixture numbers are assigned.

138 indoor occupants + 30 = 168 occupancy load design

Number of male and female occupants: 168 occupants × 50% = 84 female occupants

84 male occupants

Number of water closets: 1 water closet per 75 female occupants

$$84 \text{ females} \times \frac{1}{75} = \frac{84}{75} = 1.12 \text{ or } 2$$

females = 2 water closets

1 water closet per 75 male occupants

$$84 \text{ males} \times \frac{1}{75} = \frac{84}{75} = 1.12 \text{ or } 2$$

males = 2 water closets

Reader's note: When computing fixture numbers any remainder (fraction) of the quantity requires an additional fixture. Further 1 of the water closets may be substituted for a urinal.

Number of lavatories: 1 lavatory per 200 occupants

$$84 \text{ females} \times \frac{1}{200} = \frac{84}{200} = .42 \text{ or } 1$$

females = 1 lavatory

$$84 \text{ males} \times \frac{1}{200} = \frac{84}{200} = .42 \text{ or } 1$$

males = 1 lavatory

Note: When computing fixture numbers any remainder (fraction) of the quantity requires an additional fixture, which is similar to the water closets referenced above.

Building fixture considerations: A drinking fountain will not be required. Section 410.1 allows the omission of the drinking fountain because water is served in the restaurant. The restrooms may be designed with floor drains for overflow conditions and would be classified as emergency floor drains. These floor drains would require trap seal primers in accordance with section 1002.4 of the code.

Minimum Number of Required Plumbing Fixtures, Assembly seating

Building code occupancy numbers were commonly established to address means of exiting considerations. The plumbing code, in using the inside number of occupants, may not have ensured an adequate number of plumbing fixtures. Additionally, it is helpful to understand that the building code occupancy numbers are established to deal with fire code concerns for means of egress. The outside occupants must be factored into total occupants in order to provide sufficient restroom fixture numbers.

CHANGE TYPE. Clarification

CHANGE SUMMARY. The exception for separate facilities, number 1, has been revised to clarify an undefined term.

2006 CODE: 403.2 Separate Facilities. Where plumbing fixtures are required, separate facilities shall be provided for each sex.

403.2 continues

403.2
Separate Facilities (Dwelling Units) (Toilet Facilities)

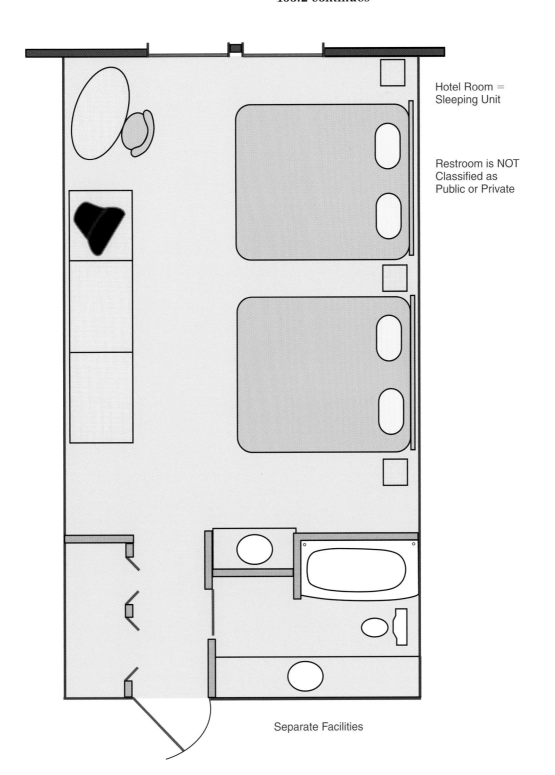

Hotel Room = Sleeping Unit

Restroom is NOT Classified as Public or Private

Separate Facilities

403.2 continued

Exceptions:

1. Separate facilities shall not be required for ~~private facilities~~ <u>dwelling units and sleeping units</u>.

*Reader's Note: *Another, larger change to this section and others has been made by the code change process and is explained in the following significant change item.*

CHANGE SIGNIFICANCE.   The code did not define *private facilities* for consideration as applying to separate facilities for each sex. Often, confusion resulted in the misinterpretation that the private facilities were to be for one person only, such as a physician's office restroom for the doctor only.

This clarification will point out that individual dwelling units and sleeping units (such as motels or hotels) will not require separate restrooms.

CHANGE TYPE.   Modifications

CHANGE SUMMARY.   The revisions use former code requirements reorganized in a manner to provide greater clarification for public and employee restrooms. The change also simplifies the provision that employee and public restrooms may be combined.

2006 CODE: 403.2 Separate Facilities.   Where plumbing fixtures are required, separate facilities shall be provided for each sex.

Exceptions:
1. Separate facilities shall not be required for <u>dwelling units and sleeping units.</u>
2. ~~Separate employee facilities shall not be required in occupancies in which 15 or less people are employed.~~
3. 2. Separate facilities shall not be required in structures or tenant spaces with a total occupant load, including both employees and customers, of 15 or less.
4. 3. Separate facilities shall not be required in mercantile occupancies in which the maximum occupant load is 50 or less.

403.2, 403.4, 403.4.1, and 403.4.2 continues

403.2, 403.4, 403.4.1, and 403.4.2

Separate Facilities, Public Facilities, Location of Toilet Facilities in Occupancies Other Than Covered Malls, and Location of Toilet Facilities in Covered Malls

Customer Access to Toilet Rooms (Combined)

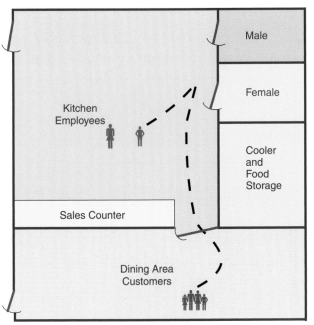

Not Acceptable

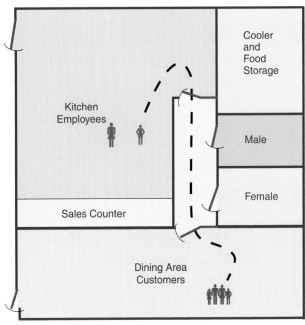

Acceptable

Location of Toilet Facilities

403.2, 403.4, 403.4.1, and 403.4.2
continued

403.4 Location of Employee Toilet Facilities in Occupancies Other Than Assembly or Mercantile. ~~Access to toilet facilities in occupancies other than mercantile and assembly occupancies shall be from within the employees' working area. Employee facilities shall be either separate facilities or combined employee and public customer facilities.~~

~~**Exception:**~~
~~Facilities that are required for employees in storage structures or kiosks, and are located in adjacent structures under the same ownership, lease or control, shall be a maximum travel distance of 500 feet (152 m) from the employees' working area.~~

~~**403.4.1 Travel Distance.** The required toilet facilities in occupancies other than assembly or mercantile shall be located not more than one story above or below the employee's working area and the path of travel to such facilities shall not exceed a distance of 500 feet (152 m).~~

~~**Exception:**~~
~~The location and maximum travel distances to required employee toilet facilities in factory and industrial occupancies are permitted to exceed that required in Section 403.4.1, provided the location and maximum travel distance are approved by the code official.~~

~~**403.5 Location of Employee Toilet Facilities in Mercantile and Assembly Occupancies.** Employees shall be provided with toilet facilities in building and tenant spaces utilized as restaurants, night~~

Customer Access to Toilet Rooms (Combined)

Mercantile Retail Store

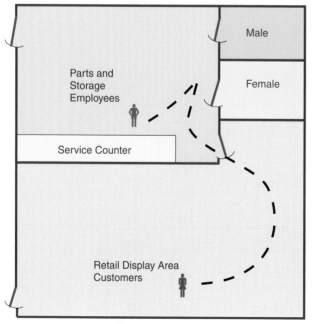

Not Acceptable

Mercantile Retail Store

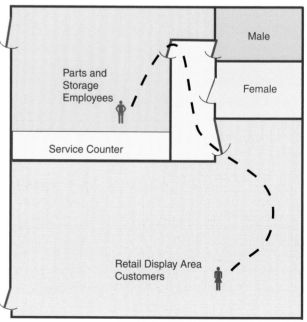

Acceptable

Location of Toilet Facilities

~~clubs, places of public assembly and mercantile occupancies. The employees' facilities shall be either separate facilities or combined employee and public customer facilities. The required toilet facilities shall be located not more than one story above or below the employee's regular work area and the path of travel to such facilities, in other than covered malls, shall not exceed a distance of 500 feet (152 m). The path of travel to required facilities in covered malls shall not exceed a distance of 300 feet (91 440 mm).~~

~~Exception:~~
~~Employee toilet facilities shall not be required in tenant spaces where the travel distance from the main entrance of the tenant space to a central toilet area does not exceed 300 feet (91 440 mm) and such central toilet facilities are located not more than one story above or below the tenant space.~~

403.6 403.4 Required Public **Toilet Facilities.** Customers, patrons, and visitors shall be provided with public toilet facilities in structures and tenant spaces intended for public utilization. The accessible route to public facilities shall not pass through kitchens, storage rooms, closets, or similar spaces. ~~Public toilet facilities shall be located not more than one story above or below the space required to be provided with public toilet facilities and the path of travel to such facilities shall not exceed a distance of 500 feet (152 m).~~ Employees shall be provided with toilet facilities in all occupancies. Employee toilet facilities shall be either separate or combined employee and public toilet facilities.

403.4.1 Location of Toilet Facilities in Occupancies Other Than Covered Malls. In occupancies other than covered malls, the required public and employee toilet facilities shall be located not more than one story above or below the space required to be provided with toilet facilities, and the path of travel to such facilities shall not exceed a distance of 500 feet (152 m).

Exception:
The location and maximum travel distances to required employee facilities in factory and industrial occupancies are permitted to exceed that required by this section, provided that the location and maximum travel distance are approved.

403. 6.1 403.4.2 Location of Toilet Facilities in Covered Malls. In covered mall buildings, the required public and employee toilet facilities shall be located not more than one story above or below the space required to be provided with toilet facilities, and the path of travel to such facilities shall not exceed a distance of 300 feet (91 440 mm). In covered mall buildings, the required facilities shall be based on total square footage, and facilities shall be installed in each individual store or in a central toilet area located in accordance with this section. The maximum travel distance to the central toilet facilities in

403.2, 403.4, 403.4.1, and 403.4.2 continues

403.2, 403.4, 403.4.1, and 403.4.2
continued

covered mall buildings shall be measured from the main entrance of any store or tenant space. In covered mall buildings, where employees' toilet facilities are not provided in the individual store, the maximum travel distance shall be measured from employee's work area of the store or tenant space.

CHANGE SIGNIFICANCE. This modification basically simplifies and clarifies the current code text through reorganization efforts.

Section 403.4 clarifies that public facilities are required. It further states that they may be combined with employee restrooms or separate from employee restrooms. The access to public facilities shall not pass through private areas such as kitchens or storage areas, to improve sanitation and security for the visitors passing through to restrooms.

Section 403.4.2 states the requirements for toilet facilities in covered malls. Past code requirements for the location of restrooms at every other floor level are clarified, and travel distances for employees are clarified.

Section 403.4.1 addresses toilet locations in occupancies other than covered malls. The requirements of restrooms allowed at every other floor level, travel distances of 500 feet, and exceptions for factory distances are previous, familiar requirements reorganized into this section to provide greater clarification.

405.3.1

Water Closets, Urinals, Lavatories, or Bidets (Clearances)

CHANGE TYPE.   Clarification

CHANGE SUMMARY.   Fixtures required by the code shall be installed with proper clearances to ensure acceptable user operation and meet sanitation requirements. The section was revised to provide editorial clarification for the referenced fixtures.

2006 CODE: 405.3.1 Water Closets, Urinals, Lavatories, or Bidets.   A water closet, urinal, lavatory, or bidet shall not be set closer than 15 inches (381 mm) from its center to any side wall, partition, vanity or other obstruction, or closer than 30 inches (762 mm) center-to-center between ~~water closets, urinals or~~ adjacent fixtures. There shall be at least a 21-inch (533 mm) clearance in front of the water closet, urinal, <u>lavatory,</u> or bidet to any wall, fixture, or door. Water closet compartments shall not be less than 30 inches (762 mm) wide and 60 inches (1524 mm) deep (see Figure 405.3.1). ~~There shall be at least a 21-inch (533 mm) clearance in front of a lavatory to any wall, fixture or door.~~

CHANGE SIGNIFICANCE.   The code change text clarifies the original intent for clearance requirements for all plumbing fixtures by reorganizing the section. The proposal does not add or delete any requirements. Removing the words *water closets* and *urinals* broadens the application to all fixtures. Further, placing the word *lavatory* in the second sentence eliminates the need for the entire last sentence.

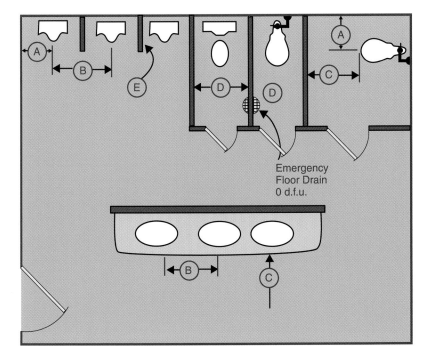

Emergency
Floor Drain
0 d.f.u.

Index

A - All fixtures shall not be closer than 15″ from center to sidewall.

B - All fixtures shall not be closer than 30″ from center to center between adjacent fixtures.

C - All fixtures shall have at least 21″ of clearance in front.

D - Water closet compartments shall not be less than 30″ wide and 60″ deep.

E - Urinal partitions shall extend at least 18″ from the wall or 6″ out from the lip.

Fixture Clearances

406.3

Waste Connection (Automatic Washer)

CHANGE TYPE.   Clarification

CHANGE SUMMARY.   The section was revised to recognize automatic washers, which drain by gravity rather than pumping action only.

2006 CODE: 406.3 Waste Connection.   The waste from an automatic clothes washer shall discharge through an air break into a standpipe in accordance with Section 802.4 or into a laundry sink. The trap and fixture drain for an automatic clothes washer standpipe shall be a minimum of 2 inches (51 mm) in diameter. The automatic clothes washer fixture drain shall connect to a branch drain or drainage stack a minimum of 3 inches (76 mm) in diameter. <u>Automatic clothes washers that discharge by gravity shall be permitted to drain to a waste receptor or an approved trench drain.</u>

CHANGE SIGNIFICANCE.   Many commercial washing machines discharge by gravity. Gravity discharge automatic washers cannot elevate the discharge up to a standpipe opening. The revised text now includes the gravity acceptance and outlines how the discharge is conveyed to the drainage system.

Waste Connection
Automatic Washers

CHANGE TYPE.   Addition

CHANGE SUMMARY.   A section was added to provide a bidet user protection from scalding.

2006 CODE: **408.3 Bidet Water Temperature.**   The discharge water temperature from a bidet fitting shall be limited to a maximum temperature of 110° F (43° C) by a water temperature limiting device conforming to ASSE 1070.

CHANGE SIGNIFICANCE.   The new section provides additional safety to reduce the risk of scalding from bidets, which had unregulated temperature protection. Protection is ensured by an ASSE 1070 device, now referenced in Chapter 13 of the code as ASSE 1070-04 Performance Requirements for Water Temperature Limiting Devices. They are also used for protection at public hand-washing facilities.

408.3
Bidet Water Temperature

Bidet Plumbing Fixture

Temperature – Limited to a maximum of 110° F

Controlled by - ASSE 1070-04, Performance Requirements for Water-temperature Limiting Devices

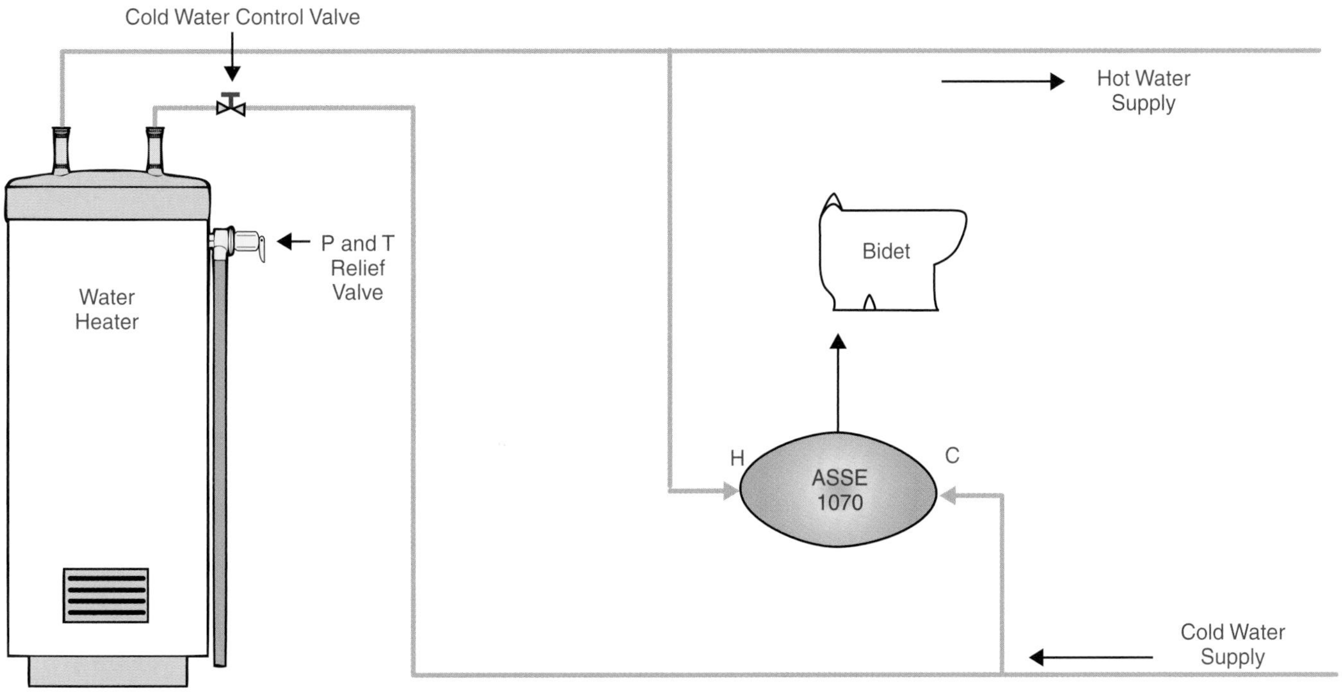

Cold Water Control Valve

Hot Water Supply

P and T Relief Valve

Water Heater

Bidet

H   ASSE 1070   C

Cold Water Supply

Bidet Water Temperature

410.1

Approval (Drinking Fountains)

CHANGE TYPE.   Modification

CHANGE SUMMARY.   The revised text now accepts water cooler or bottled water dispensers as providing drinking water for up to half of the required fountains.

2006 CODE: 410.1 Approval.   Drinking fountains shall conform to ASME A112.19.1M, ASME A112.19.2M, or ASMEA112.19.9M and water coolers shall conform to ARI 1010. Drinking fountains and water coolers shall conform to NSF 61, Section 9. Where water is served in restaurants, drinking fountains shall not be required. In other occupancies, where drinking fountains are required, <u>water coolers or</u> bottled water dispensers shall be permitted to be substituted for not more than 50% of the required drinking fountains.

CHANGE SIGNIFICANCE.   Drinking water quality has often been a major consideration when discussing commercially provided water from dispensers rather than drinking fountain fixtures supplied by the building's water distribution system. Additionally, the code change was supported by many statements that "water coolers and bottled water dispensers are viable listed alternatives to drinking fountains." It was also explained that these units provide acceptable accessibility alternatives to standard accessible, compliant drinking fountains.

Clear Container

When drinking fountains are required

First - Drinking fountains conforming to the standards listed shall be provided

Then - Water coolers or bottled water dispensers may be used for up to 50% of required drinking fountains

That concern is resolved by the limit of water coolers and bottled water dispensers to 50% of the required drinking fountains.

It is important to note that the water coolers referenced above are not the commonly considered ARI 1010, Self-Contained, Mechanically-Refrigerated Drinking Water Coolers. The original proposal requested the deletion of bottled water dispensers, and the final code revision expanded the water source to include water coolers.

Readers are reminded that structures constructed to code requirements are mandated to continue and maintain the code-required fixtures during the building's use. This is of critical concern when considering arguments on both sides of this issue with regard to the cleanliness of drinking fountains and the continued availability and supply of the bottled water dispensers.

412.2

Floor Drain (Strainer)

CHANGE TYPE.   Clarification

CHANGE SUMMARY.   The section has been revised to clarify that the strainer surface was being considered rather than the trap drain area or measurement.

2006 CODE: 412.2 Floor Drains ~~Trap and Strainer.~~   Floor drains ~~traps~~ shall have removable strainers. ~~The strainer shall have a waterway area of not less than the area of the tailpiece.~~ The floor drain shall be constructed so that the drain is capable of being cleaned. Access shall be provided to the drain inlet.

CHANGE SIGNIFICANCE.   The revision clarifies that the discussion centers on the strainer, not the trap. The opening area of a strainer (free area) is specified in standard ASME A112.6.3, found in section 412.1. The standard addresses free area of 48% to 69% greater than the drain area.

Floor drain strainers, traps, and line sizes are guided by basic plumbing code principals. Section 704.2 states "the size of a drainage pipe shall not be reduced in size in the direction of flow."

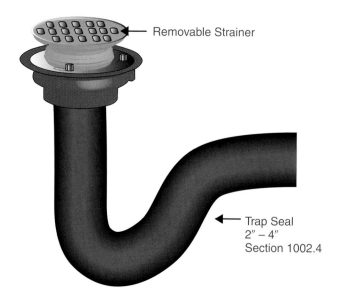

Removable Strainer

Trap Seal
2″ – 4″
Section 1002.4

Floor Drains

CHANGE TYPE. Modification

CHANGE SUMMARY. The section has been revised to include an exception. The exception to the normal square shower floor area now includes a rectangular floor area, which is that of a conventional bathtub.

2006 CODE: 417.4 Shower Compartments. All shower compartments shall have a minimum of 900 square inches (0.58 m^2) of interior cross-sectional area. Shower compartments shall not be less than 30 inches (762 mm) in minimum dimension measured from the finished interior dimension of the compartment, exclusive of fixture valves, showerheads, soap dishes, and safety grab bars or rails. Except as required in Section 404, the minimum required area and dimension shall be measured from the finished interior dimension at a height equal to the top of the threshold and at a point tangent to its centerline and shall be continued to a height not less than 70 inches (1778 mm) above the shower drain outlet.

Exception:

Shower compartments having not less than 25 inches (635 mm) in minimum dimension measured from the finished interior dimension of the compartment, provided the shower compartment has a minimum of 1300 square inches (.838 m^2) of cross-sectional area.

CHANGE SIGNIFICANCE. The 2003 edition of the code implied that the minimum acceptable floor area of a shower stall would be a 30" square opening. This code change considers a rectangular floor space normally designed in a structure for a bathtub. This exception allows a rectangular shower stall to be accepted and installed in that tub

417.4 continues

417.4
Shower Compartments (Floor Area)

Shower Compartment

417.4 continued

space area without making major wall changes to the structure. The new text allows a reduction in the width from 30 inches to 25 inches; however, the length is extended. Code users are reminded that the regulations provide the minimum qualifications, and the practical application is assumed to be a 30-inch by 60-inch shower basin.

A manufacturer's representative submitted the code change, understanding the need for replacement of bathtubs with similar size showers. The replacement is often necessary for an aging population to have greater access to the shower. This greater access issue is understandable when considering the vast majority of individuals use the tub and shower fixture for showering. Conversion of a tub and shower to a shower eliminates the higher threshold in which a less mobile person would have to step over.

417.4.2

Access (to Showers)

CHANGE TYPE.   Addition

CHANGE SUMMARY.   Code information has been added to identify the minimum size of 22 inches for access to a shower.

2006 CODE: <u>**417.4.2 Access.** The shower compartment access and egress opening shall have a minimum clear and unobstructed finished width of 22 inches (559 mm). Shower compartments required to be designed in conformance to accessibility provisions shall comply with Section 404.1.</u>

CHANGE SIGNIFICANCE.   Previously the code did not provide a minimum width for shower compartment access and egress openings. Without a clearly stated minimum width, code officials and installers were unable to uniformly apply the access opening so that the shower compartment is functionally accessible.

The new code section will clarify the opening size and improve other considerations. Those considerations are an ensured entrance and exit opening for the user; improved access for cleaning; improved access for maintenance to the valve and drain; and acceptable opening for emergency response when necessary.

The 22-inch provision is based upon earlier building code commentary discussion of the measurement of egress consideration for an approximation of average shoulder width of an adult.

Access (To Showers)

419.1
Approval

Approval (Urinals)

CHANGE TYPE.   Modification

CHANGE SUMMARY.   The ANSI Z124.9 standard is now included in the code section with other design and performance standards previously recognized. This standard will provide recognition and acceptance of plastic waterless urinals. The section is also revised to clarify that not all urinals depend on trap siphonage generated by a water supply.

2006 CODE: 419.1 Approval.   Urinals shall conform to <u>ANSI Z124.9,</u> ASME A112.19.2M, CSA B45.1, or CSA B45.5. Urinals shall conform to the water consumption requirements of Section 604.4. <u>Water supplied</u> ~~Urinals~~ <u>urinals</u> shall conform to the hydraulic performance requirements of ASME A112.19.6, CSA B45.1, or CSA B45.5.

CHANGE SIGNIFICANCE.   This change was submitted to add a standard for plastic urinals with performance requirements. ANSI Z124.9, the standard for material requirements for plastic urinals, includes waterless urinals.

A previous code change added ANSI Z124.9, a standard for plastic waterless urinals. The latter part of the 2003 code section assumed that all urinals had a water supply. Confusion would have resulted. This change clarifies that only urinals with a water supply should conform to the hydraulic performance requirements of ASME A112.19.6, CSA B45.1, or CSA B45.5.

To date, many jurisdictions having authority have individually been required to make acceptance decisions for the waterless urinals. Water conservation, proper sanitation, and maintenance issues have always been part of the acceptance issues. The code now allows acceptance based upon the international code process for inclusion of national consensus standards.

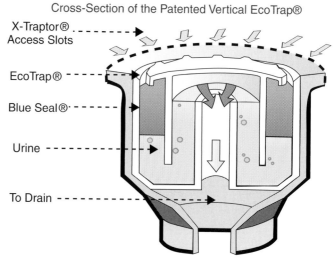

Cross-Section of the Patented Vertical EcoTrap®

X-Traptor® Access Slots

EcoTrap®

Blue Seal®

Urine

To Drain

Approval (Urinals)

CHANGE TYPE.   Modification

CHANGE SUMMARY.   The text clarifies that assembly and educational use groups may have up to 67% of the required number of water closet fixtures converted to urinals. The revision further explains that all other occupancies shall not have urinals substituted for more than 50% of required water closets.

2006 CODE: 419.2 Substitution for Water Closets.   In each bathroom or toilet room, urinals shall not be substituted for more than 67% of the required water closets <u>in assembly and educational occupancies. Urinals shall not be substituted for more than 50% of the required water closets in all other occupancies.</u>

CHANGE SIGNIFICANCE.   The 2003 code edition allowed the substitution of urinals for 67% (²/₃) of the water closets located in all use groups that included assembly and educational occupancies. The code change continues the concept for assembly and educational classifications.

   This approved proposal mandates that all other use groups may have up to only half the number of water closets substituted as urinals. Prior to the publication of the International Plumbing Code, several codes limited urinal substitutions to 50%.

419.2

Substitution for Water Closets (Urinals)

Assembly and Educational Occupancies

   67% Maximum allowable substitution of urinals for water closets.

All Other Occupancies

   50% Maximum allowable substitution of urinals for water closets.

421.2 and 421.5

Installation and Access to Pump (Whirlpool Tubs)

CHANGE TYPE.   Clarification

CHANGE SUMMARY.   Greater detail is added in the code to ensure access to tub pump assemblies. The change provides opening size criteria and includes the manufacturer's installation instructions.

2006 CODE: 421.2 Installation.   Whirlpool bathtubs shall be installed and tested in accordance with the manufacturer's installation instructions. The pump shall be located above the weir of the fixture trap. ~~Access shall be provided to the pump.~~

421.5 Access to Pump.   Access shall be provided to circulation pumps in accordance with the fixture or pump manufacturer's installation instructions. Where manufacturer's instructions do not specify the location and minimum size of field-fabricated access openings, a 12-inch by 12-inch (305 mm by 305 mm) minimum sized opening shall be installed to provide access to the circulation pump. Where pumps are located more than 2 feet (609 mm) from the access opening, an 18-inch by 18-inch (457 mm by 457 mm) minimum size opening shall be installed. A door or panel shall be permitted to close the opening. In all cases, the access shall be unobstructed and of the size necessary to permit the removal and replacement of the circulation pump.

CHANGE SIGNIFICANCE.   The code required access to the whirlpool pumps but did not provide clarification and minimum clearance in-

Access Built into Unit at Factory

formation. Several committee members felt that manufacturers addressed the requirements in their installation instructions. Yet others stated not all manufacturers provided the information, and often the fixtures were installed in a manner that differed from that described in the installation instructions, such as wall or platform configurations.

Code officials, installers, and owners understand the extremely high cost to future owners when a repair is required. Access to pumps has commonly been overlooked and incurs unnecessary cost, labor,

421.2 and 421.5 continues

Installation and Access to Pump

Not Acceptable Due to Water Line Obstructions

421.2 and 421.5 continued   and material for making them accessible. Maintenance and repairs after occupancy may involve costly repairs to ceramic tile walls or even ceilings simply to change a gasket or tighten a union.

The code change allows manufacturers to provide proper guidance but it also adds minimum standards where the manufacturer's provisions are not complete. The new text addresses another factor, that piping cannot extend across the access opening.

The P27-04/05 code change was accepted, which further revised the section to focus on the opening rather than the door or panel. The term *pump* was added in place of *fixture* since the pump may have requirements that differ from those of the fixture manufacturer.

CHANGE TYPE.   Modification

CHANGE SUMMARY.   The revised section lists two new standards, which specifically address waste fittings.

2006 CODE: 424.1.2 Waste Fittings.   Waste fittings shall conform to <u>ASME A112.18.2, ASTM F409, CSA B125,</u> or to one of the standards listed in Tables 702.1 and 702.4 for above-ground drainage and vent pipe and fittings.~~, or the waste fittings shall be constructed of tubular stainless steel with a minimum wall thickness of 0.012 inch (0.30 mm), tubular copper alloy having a minimum wall thickness of 0.027 inch (0.69 mm) or tubular plastic complying with ASTM F409.~~

Chapter 13 Referenced Standards.   <u>ASME A112.18.2–02 Plumbing Fixture Waste Fittings 424.1.2.</u>

CHANGE SIGNIFICANCE.   The changed text provides new referenced standards ASME A 112.18.2 and CSA B125, developed as consensus standards which address plumbing fixture waste fittings. The deleted material specifications and dimensions are now covered in the ASME and CSA standards.

424.1.2
Waste Fittings

Waste Fittings

Waste Fittings

424.2

Hand Showers

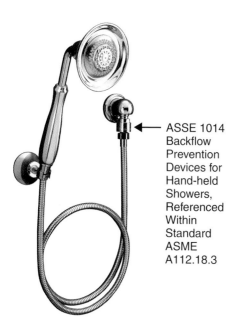

← ASSE 1014 Backflow Prevention Devices for Hand-held Showers, Referenced Within Standard ASME A112.18.3

CHANGE TYPE. Modification

CHANGE SUMMARY. The section now contains the most recent standard, which addresses backflow protection.

2006 CODE: 424.2 Hand Showers. Hand-held showers shall conform to ~~ASSE 1014,~~ ASME A112.18.1 or CSA B125.1. Hand-held showers shall provide backflow protection in accordance with ASME A112.18.1 or CSA B125.1 or shall be protected against backflow by a device complying with ASME A112.18.3.

CHANGE SIGNIFICANCE. The section change provides the correct standards reference for hand-held showers to ensure backflow protection. Hand showers similar to faucets and accessories are fixture fittings, which are regulated by ASME A112.18.1 and CSA B125. These standards address all product performance, including mechanical wear, temperature and pressure resistance, and surface finishing, in addition to backflow prevention. ASSE 1014 is referenced within ASME A112.18.3 and remains referenced in the code with other backflow protection devices.

ASSE Standard #1014-2005
ASSE Board Approved: JANUARY, 2005
ANSI APPROVED: APRIL, 2005

American Society of Sanitary Engineering

Performance Requirements for
Backflow Prevention Devices for Hand-Held Shower

An American National Standard

424.3 and 424.4

Individual Shower Valves and Multiple (Gang) Showers Supplied with a Single Tempered Water Supply Pipe

CHANGE TYPE. Modification

CHANGE SUMMARY. Section 424.3 has been revised to identify individual showers only and prohibit individual shower and tub-shower combination valves from having temperature protection provided by in-line thermostatic valves. Section 424.4 has been added to address gang showers with multiple valves for several occupants. The revised section resulting from change P29-04/05 requires that temperature be protected by a device conforming to ASSE 1069.

2006 CODE: **424.3 Individual Shower Valves.** Individual shower and tub-shower combination valves shall be balanced-pressure, thermostatic or combination balanced-pressure/thermostatic valves that conform to the requirements of ASSE 1016 or CSA B125 and shall be installed at the point of use. Multiple (gang) showers supplied with a single tempered water supply pipe shall have the water supply for such showers controlled by a master thermostatic mixing valve complying with ASSE 1017. Shower and tub-shower combination valves and master thermostatic mixing valves required by this section shall be equipped with a means to limit the maximum setting of the valve to 120° F (49° C), which shall be field-adjusted in accordance with the manufacturer's instructions. In-line thermostatic valves shall not be utilized for compliance with this section.

424.3 and 424.4 continues

Temperature – Limited to a maximum of 120° F

Controlled by - ASSE 1016-96, Performance Requirements for Individual Thermostatic, Pressure Balancing and Combination Control Valves for Individual Fixture Fittings

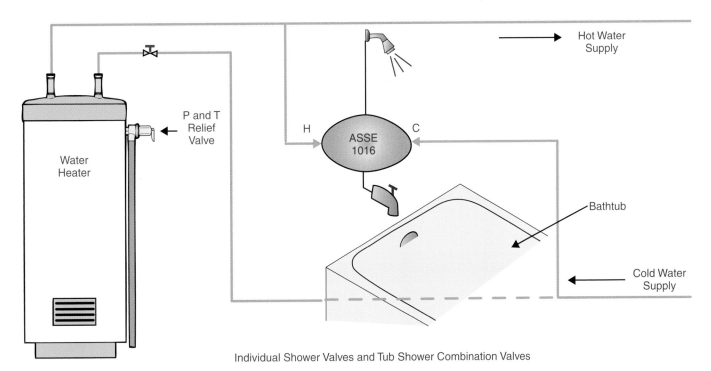

Individual Shower Valves and Tub Shower Combination Valves

424.3 and 424.4 continued

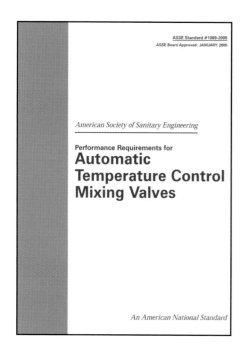

ASSE Standard #1069-2005
ASSE Board Approved: JANUARY, 2005

American Society of Sanitary Engineering

Performance Requirements for
Automatic Temperature Control Mixing Valves

An American National Standard

424.4 Multiple (Gang) Showers. Multiple (gang) showers supplied with a single tempered water supply pipe shall have the water supply for such showers controlled by an approved automatic temperature control mixing valve that conforms to ASSE 1069 or CSA B125, or each shower head shall be individually controlled by a balanced pressure, thermostatic or combination balanced-pressure/thermostatic valve that conforms to ASSE 1016 or CSA B125 and shall be installed at the point of use. Such valves shall be equipped with a means to limit the maximum setting of the valve to 120° F (49° C), shall be field-adjusted in accordance with the manufacturer's instructions.

Add new standard to Chapter 13 as follows:
ASSE 1069–05 Performance Requirements for Automatic Temperature Control Mixing Valves 424.4

CHANGE SIGNIFICANCE. The revision to section 424.3 clarifies that where individual shower valves are installed, a single tempered water supply pipe must be controlled by an approved mixing valve, an ASSE 1016 standard device. Section 424.3 was further revised to remove multiple (gang) showers into a section of their own. A code change provided that in-line thermostatic valves shall not be accepted for providing temperature protection. The proposal advocate stated in-line thermostatic valves do not provide for thermal shock protection for individual shower applications since there is further mixing downstream, which negates the regulation of the temperature required

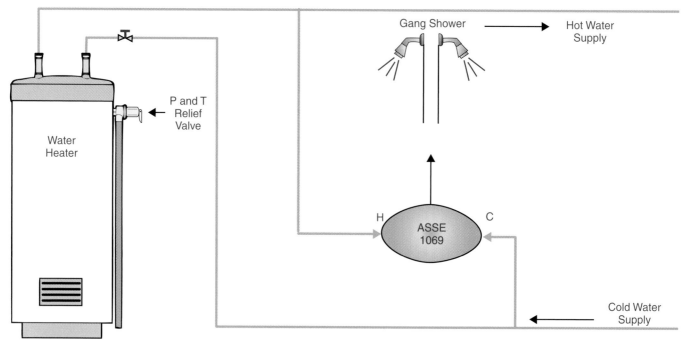

Temperature – Limited to a maximum of 120° F

Controlled by - ASSE 1069-05, Performance Requirements for Automatic Temperature Control Mixing Valves

Multiple (Gang) Showers

by the ASSE 1016. If the cold water supply to a mixing valve is unprotected, which occurs with in-line devices installed only on the hot water supply, there will be temperature variations in the outlet of the shower head above those required in the ASSE 1016 standard. Thermal shock can result in situations where the occupant may be startled, resulting in slip-and-fall injuries.

The added section 424.4, Multiple (Gang) Showers, provides acceptance information for the temperature protection for multiple control valves supplied by a single automatic temperature control for several occupants. Later, the revised section includes a referenced standard for temperature control on multiple (gang) showers, ASSE 1069-05, Performance Requirements for Automatic Temperature Control Mixing. The valve was developed to address gang shower protection and is installed in-line to supply multiple shower stations with a single water supply line.

ASSE 1016 or CSA B125 is included for temperature control where each gang shower is individually protected. The choice may be made by the designer or the installer.

424.5

Bathtub and Whirlpool Bathtub Valves

Bathtub and Whirlpool Bathtub Valves

CHANGE TYPE. Addition

CHANGE SUMMARY. A section has been added that now requires bathtubs and whirlpool bathtubs to have temperature protection provided by an ASSE 1070 device.

2006 CODE: <u>**424.5 Bathtub and Whirlpool Bathtub Valves.** The hot water supplied to bathtubs and whirlpool bathtubs shall be limited to a maximum temperature of 120° F (49° C) by a water temperature limiting device that conforms to ASSE 1070, except where such protection is otherwise provided by a combination tub/shower valve in accordance with Section 424.3.</u>

Chapter 13. ASSE 1070-04 Performance Requirements for Water Temperature Limiting Devices 424.5

CHANGE SIGNIFICANCE. Manufacturers, code officials, and the design community have discussed adding scald protection for several code cycles. Research and improved protection products have motivated the code community to consider additional area protection, such as for tubs, rather than just slipping issues in showers.

This new section provides protection from scalding in bathtubs and whirlpool bathtubs by requiring a water-temperature-limiting device that conforms to an approved referenced standard, ASSE

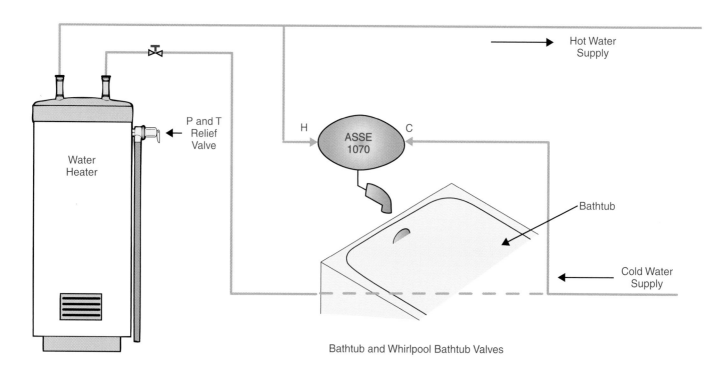

Temperature – Limited to a maximum of 120° F

Controlled by - ASSE 1070-04, Performance Requirements for Water Temperature Limiting Device

Bathtub and Whirlpool Bathtub Valves

1070-04. This standard provides for devices to be installed with the fixture fitting or they can be integral to the plumbing fixture fitting supplying the water.

A further modification in the code change process added whirlpool bathtubs to the title and code text. Temperature control limitations should cover installations for both bathtub and whirlpool bathtub valves.

504.6

Requirements for Discharge Piping (Structural Damage)

CHANGE TYPE.   Modification

CHANGE SUMMARY.   Three sections were deleted and a new section was created to clarify the previous code requirement for relief valve discharge lines. Item 6 was changed to clarify damage resulting from relief valve discharge.

2006 CODE: ~~**504.6 Relief Outlet Waste.**   The outlet of a pressure, temperature or other relief valve shall not be directly connected to the drainage system.~~

~~**504.6.1 Discharge.**   The relief valve shall discharge full-size to a safe place of disposal such as the floor, outside the building, or an indirect waste receptor. The discharge pipe shall not have any trapped sections and shall have a visible air gap or air gap fitting located in the same room as the water heater. The outlet end of the discharge pipe shall not be threaded and such discharge pipe shall not have a valve or tee installed. Relief valve piping shall be piped independent of other equipment drains or relief valve discharge piping to the disposal point. Such pipe shall be installed in a manner that does not cause personal injury to occupants in the immediate area or structural damage to the building.~~

~~**504.6.2 Materials.**   Relief valve discharge piping shall be of those materials listed in Section 605.4 or shall be tested, rated and approved for such use in accordance with ASME A112.4.1. Piping from safety pan drains shall be of those materials listed in Table 605.4.~~

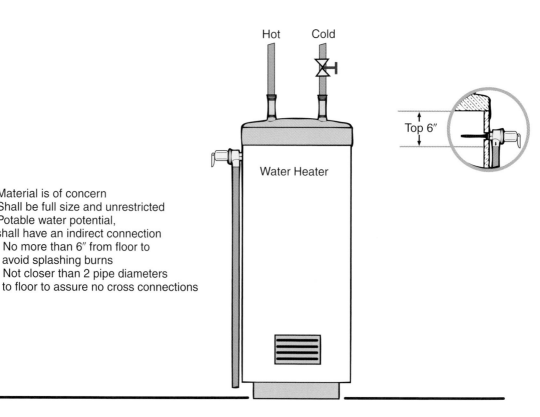

- Material is of concern
- Shall be full size and unrestricted
- Potable water potential,
  shall have an indirect connection
  - No more than 6″ from floor to
    avoid splashing burns
  - Not closer than 2 pipe diameters
    to floor to assure no cross connections

Hot   Cold

Water Heater

Top 6″

504.6 Requirements for Discharge Piping. The discharge piping serving a pressure relief valve, temperature relief valve, or combination thereof shall:

1. Not be directly connected to the drainage system.

2. Discharge through an air gap located in the same room as the water heater.

3. Not be smaller than the diameter of the outlet of the valve served and shall discharge full size to the air gap.

4. Serve a single relief device and shall not connect to piping serving any other relief device or equipment.

5. Discharge to the floor, an indirect waste receptor, or to the outdoors. Where discharging to the outdoors in areas subject to freezing, discharge piping shall be first piped to an indirect waste receptor through an air gap located in a conditioned area.

6. Discharge in a manner that does not cause personal injury or ~~property~~ structural damage.

7. Discharge to a termination point that is readily observable by the building occupants.

8. Not be trapped.

9. Be installed so as to flow by gravity.

10. Not terminate more than 6 inches (152 mm) above the floor or waste receptor.

11. Not have a threaded connection at the end of such piping.

12. Not have valves or tee fittings.

13. Be constructed of those materials listed in Section 605.4 or materials tested, rated, and approved for such use in accordance with ASME A112.4.1.

Section 504.4 Combination pressure and temperature relief valve conforming to ANSI Z21.22

Section 504.4.1 shall be installed in the shell of the water heater tank

Requirements for Discharge Piping

CHANGE SIGNIFICANCE. The coded change created a new format to address all discharge piping requirements in a list rather than 13 re-

504.6 continues

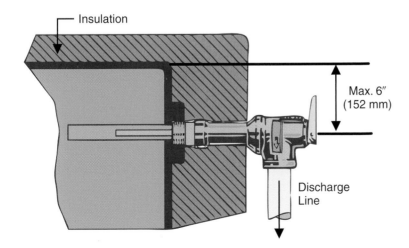

504.6 continued quirements in three different paragraphs. The individual proposing the change stated that the past method of presentation was too cumbersome, easily misunderstood, and often misapplied. The requirements were taken from existing information.

Item 6 was revised and improved to address structural damage resulting from relief valve discharge rather than property damage by another approved code change. The 2003 edition of the *International Plumbing Code* used the term *structural damage,* not *property damage.* Property damage is not defined in the code and can be construed to include not only building components (e.g., flooring, drywall, or baseboards) but personal property as well. Relief line discharge, similar to fire sprinkler activation, is an emergency response to a dangerous situation. When the discharge takes place, the code should protect against structural damage, not property damage, which will vary according to building usage.

CHANGE TYPE. Clarification

CHANGE SUMMARY. The sections are revised to place the piping requirements in the correct location.

2006 CODE: 504.7.1 Pan Size and Drain. The pan shall be not less than 1.5 inches (38 mm) deep and shall be of sufficient size and shape to receive all dripping or condensate from the tank or water heater. The pan shall be drained by an indirect waste pipe having a minimum diameter of ¾ inch (19 mm). <u>Piping for safety pan drains shall be of those materials listed in Table 605.4.</u>

CHANGE SIGNIFICANCE. The 2003 text that clarified which drain piping could be used in conjunction with which pans was located in

504.7.1 continues

504.7.1
Pan Size and Drain

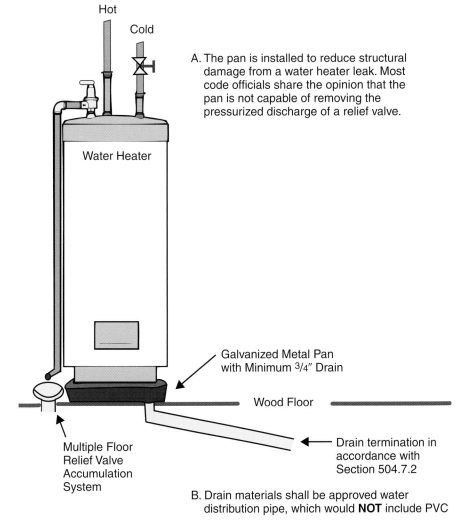

Hot
Cold

A. The pan is installed to reduce structural damage from a water heater leak. Most code officials share the opinion that the pan is not capable of removing the pressurized discharge of a relief valve.

Water Heater

Galvanized Metal Pan with Minimum 3/4″ Drain

Wood Floor

Multiple Floor Relief Valve Accumulation System

Drain termination in accordance with Section 504.7.2

B. Drain materials shall be approved water distribution pipe, which would **NOT** include PVC

Note: Items A and B are controversial issues and must be resolved during the structures design.

Pan Size and Drain

504.7.1 continued the incorrect section (504.6.2). The text is now located correctly in Section 504.7.1 because it contains the requirements for piping materials for safety drain pans.

Code officials and installers are strongly cautioned here to consider the ramifications of material selection for drain pan piping by the contractor. The table reference 605.4 is for water distribution pipe, which considers piping for temperatures higher than common drainage waste and vent piping. PVC, which is the common piping selected by installers as drain material for smaller lines, is not included in the table. The most frequent application here would be multiple pans for water heaters draining through a stacked apartment heater installation.

604.5, 604.10, and 202

Size of Fixture Supply and Gridded and Parallel Water Distribution System Manifolds

CHANGE TYPE.   Modification

CHANGE SUMMARY.   Two sections are revised to reference a new water piping concept: gridded systems. A definition is added to Chapter 2 to define the concept, which is similar to previously accepted parallel systems.

2006 CODE: 604.5 Size of Fixture Supply.   The minimum size of a fixture supply pipe shall be as shown in Table 604.5. The fixture supply pipe shall not terminate more than 30 inches (762 mm) from the point of connection to the fixture. A reduced-size flexible water connector installed between the supply pipe and the fixture shall be of an approved type. The supply pipe shall extend to the floor or wall adjacent to the fixture. The minimum size of individual distribution lines utilized in <u>gridded or</u> parallel water distribution systems shall be as shown in Table 604.5.

604.10 <u>Gridded and</u> Parallel Water Distribution System Manifolds.   Hot water and cold water manifolds installed with <u>gridded or</u> parallel connected individual distribution lines to each fixture or fixture fittings shall be designed in accordance with Sections 604.10.1 through 604.10.3.

SECTION 202 GENERAL DEFINITIONS

<u>Gridded Water Distribution System.</u>   <u>A water distribution system where every water distribution pipe is interconnected so as to provide two or more paths to each fixture supply pipe.</u>

604.5, 604.10, and 202 continues

Parallel Pex System

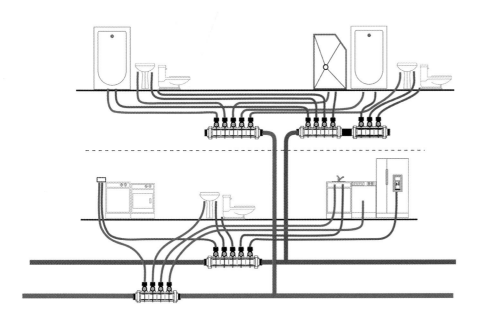

Gridded and Parallel Water Distribution Systems

604.5, 604.10, and 202 continued

CHANGE SIGNIFICANCE. Gridded water distribution systems have been added to the code as an accepted alternative to other systems used by installers

A cross-linked polyethylene (PEX) water piping manufacturer expressed his opinion that a gridded water distribution system may be more hydraulically efficient than a parallel distribution system or branch line layout. The grid also balances the pressure throughout the system, so the plumbing fixture that is the farthest from the water service may have practically the same available pressure as the one closest to the water service. This will eliminate complaints of insufficient pressure at plumbing fixtures that are distant from the water service pipe. The grid reduces the coefficient of friction loss by splitting the water into small volumes that are moving at lower velocities. At the connection to a fixture supply pipe, the water arrives from two or more directions, depending upon its location in the grid.

The manufacturer further stated that without a code clarification some plumbing officials are requiring that the plans be stamped by an engineer, increasing the installation cost for these systems. Gridded layouts are hydraulically designed by computer. The output shows the most efficient paths to each plumbing fixture and fixture fitting and documents the volume and pressure available at each fixture supply pipe. The gridded water distribution concept originated in residential sprinkler installations having PEX piping.

CHANGE TYPE. Clarification

CHANGE SUMMARY. The section was revised to clarify where a water service should terminate. In the past, confusion often developed when dealing with plastic water services rather than the acceptable material.

2006 CODE: 605.3 Water Service Pipe. Water service pipe shall conform to NSF 61 and shall conform to one of the standards listed in Table 605.3. All water service pipe or tubing, installed underground and outside of the structure, shall have a minimum working pressure rating of 160 psi (1100 kPa) at 73.4° F (23° C). Where the water pressure exceeds 160 psi (1100 kPa), piping material shall have a minimum rated working pressure equal to the highest available pressure. ~~Plastic water service piping shall terminate within 5 feet (1524mm) inside of the point where the pipe penetrates an exterior wall or slab on grade.~~ <u>Water service piping materials not third-party certified for water distribution shall terminate at or before the full open valve located at the entrance to the structure.</u> All ductile iron water service piping shall be cement mortar lined in accordance with AWWA C104.

CHANGE SIGNIFICANCE. The previous code language appeared to limit the use of an approved material (plastic water service pipe) from entering the structure past the building walls. This often resulted in installers using an unnecessary transition fitting and other piping in inaccessible areas, thereby increasing the possibility of a leak.

605.3 continues

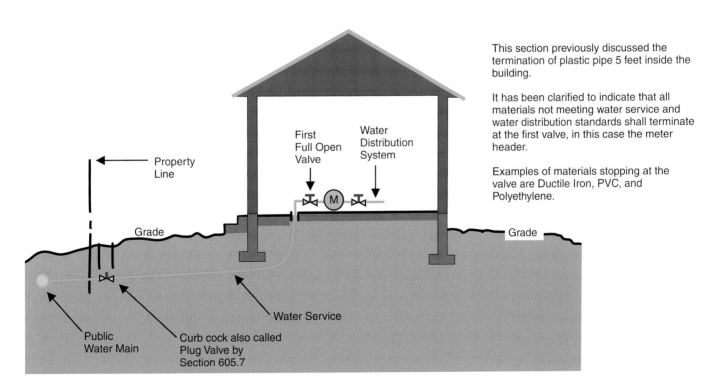

This section previously discussed the termination of plastic pipe 5 feet inside the building.

It has been clarified to indicate that all materials not meeting water service and water distribution standards shall terminate at the first valve, in this case the meter header.

Examples of materials stopping at the valve are Ductile Iron, PVC, and Polyethylene.

Water Service Pipe

605.3 continued   This code change includes a much-needed specific new line of division, where the water service piping ends and the water distribution piping starts, at or before the valve located at the entrance to the structure. The water service terminates where the water distribution starts, and this is already covered by existing definitions. The code requires a service valve at the supply pipe entrance to the structure (Section 606.1, Item 2) and allows the water service piping to remain uninterrupted until it reaches a logical place to transition, at the water distribution piping system.

CHANGE TYPE.   Additional Materials

CHANGE SUMMARY.   The code revisions updated three water service, distribution, and fitting material tables to include two new piping materials and their accompanying standards.

2006 CODE:

Table 605.3, Table 605.4, Table 605.5

Water Service Pipe, Water Distribution Pipe, and Pipe Fittings

TABLE 605.3   Water Service Pipe

Material	Standard
Polypropylene	ASTM F 2389 CSA B137.11
Cross-linked polyethylene/aluminum/high density polyethylene (PEX-AL-HDPE)	ASTM F1986

__Reader's Note:__ Other changes to these tables have been made by the code change process and are explained in additional significant change items. Further portions of tables not covered in this book as significant changes will remain as printed from the 2003 code edition.

TABLE 605.4   Water Distribution Pipe

Material	Standard
Polypropylene	ASTM F 2389 CSA B137.11
Cross-linked polyethylene/aluminum/high density polyethylene (PEX-AL-HDPE)	ASTM F1986

Table 605.3, Table 605.4, Table 605.5 continues

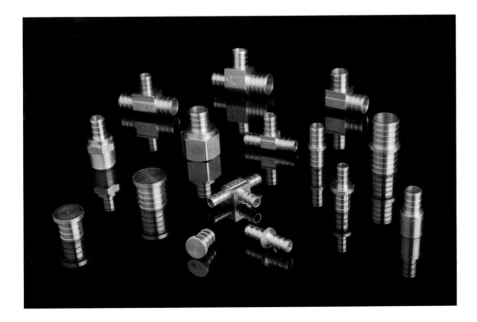

Fittings Conforming to Standard ASTM F877

Table 605.3, Table 605.4, Table 605.5 continued

TABLE 605.5 Pipe Fittings

Material	Standard
Polypropylene	ASTM F 2389
	CSA B137.11
Cross-linked polyethylene/aluminum/high density polyethylene (PEX-AL-HDPE)	ASTM F1986

CHANGE SIGNIFICANCE. The code changes, listed above, allowed the use of polypropylene in hot and cold water distribution piping. Standards ASTM F 2389 and CSA B137.11 provide acceptance criteria for inclusion in various tables of the *International Plumbing Code.* The standards require the minimum rating of 160 psi at 73° F (water service) and 100 psi at 180° F (hot and cold water distribution). Polypropylene materials meeting these requirements have over 30 years of successful history in water distribution and chemical process piping and have been used extensively in Europe. Requirements for dimensions, materials, pressure rating, performance tests, thermal stability, and compliance with NSF 61 are required.

Another code change allowed the use of cross-linked polyethylene/ aluminum/high density polyethylene with the standard acceptance criteria established in ASTM F1986. The standard addresses multilayer pipe, compression fittings, and compression joints for hot and cold water distribution systems. The code presently accepts a similar product, polyethylene/aluminum/high density polyethylene.

Table 605.5

Fittings (Water Supply and Distribution)

CHANGE TYPE.   Modification

CHANGE SUMMARY.   The table identifies which water supply and distribution fittings are acceptable and now allows ASTM F2159.

2006 CODE:

TABLE 605.5   Fittings

Material	Standard
Fittings for cross-linked polyethylene (PEX) plastic tubing	ASTM F 1807, ASTM F 1960, ASTM F 2080, <u>ASTM F 2159</u>

Reader's Note: *Other changes to the table have been made by the code change process and are explained in additional significant change items. Further portions of tables not covered in this book as significant changes will remain as printed from the 2003 code edition.*

Add new standard to Chapter 13 as follows:
<u>ASTM F2159–03 Standard Specification for Plastic Insert Fittings Utilizing a Copper Crimp Ring for SDR9 Cross-Linked Polyethylene (PEX) Tubing</u> . . . . Table 605.5

Table 605.5 continues

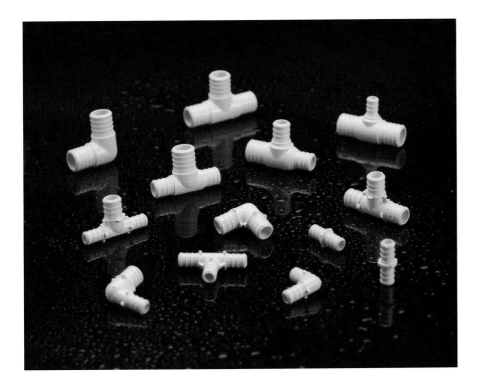

Fittings Conforming to Standard ASTM F2159

Table 605.5 continued

CHANGE SIGNIFICANCE.   The purpose of this code change is to include a nationally recognized standard for fittings used with PEX SDR9 tubing, the ASTM F2159. In the 2003 edition the full title is Standard Specification for Plastic Insert Fittings Utilizing a Copper Crimp Ring for SDR9 Cross-Linked Polyethylene (PEX) Tubing. Currently there are a number of approved fitting jointing methods; PEX pipe is not fused or solvent-cemented.

CHANGE TYPE. Modification

CHANGE SUMMARY. The section's revision clarifies which water supply and distribution valves shall meet standard NSF 61.

2006 CODE: 605.7 Valves. All valves shall be of the approved type and compatible with the type of piping material installed in the system. <u>Ball valves, gate valves, globe valves, and plug valves intended to supply drinking water shall meet the requirements of NSF 61.</u>

CHANGE SIGNIFICANCE. The 2003 *International Plumbing Code* requires conformance with NSF 61, Drinking Water System Components—Health Effects, for pipes, fittings, and faucets. However, it did not make clear that valves should be considered among these components with regard to health effects. The standard ensures that products do not leach dangerous levels of lead or other hazardous chemicals into drinking water, in this case through valves.

Plug valves have an internally ported, tapered body with a lever or square head. They are commonly called curb cocks and are located on water services. The valves are used to isolate private-property plumbing from public systems.

605.7
Valves (Water Supply)

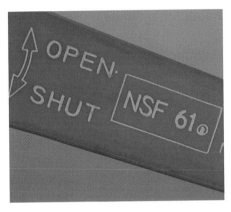

Valves

NSF/ANSI 61 – 2003e

Drinking water system components — Health effects

NSF International Standard/
American National Standard

Developed by a consortium of:
- NSF International
- The American Water Works Association Research Foundation
- The Association of State Drinking Water Administrators
- The American Water Works Association

With support from:
- The U.S. Environmental Protection Agency
  under cooperative agreement #CR-812144

NSF/ANSI 61 – 2003e

NSF®

605.17.2, Table 605.5

Mechanical Joints (for Cross-Linked Polyethylene Plastic) and Pipe Fittings

CHANGE TYPE. Modification

CHANGE SUMMARY. The section and table where revised to include a new standard that addresses the mechanical joining of cross-linked polyethylene plastic for water supply and distribution.

2006 CODE: 605.17.2 Mechanical Joints. Mechanical joints shall be installed in accordance with the manufacturer's instructions. Fittings for cross-linked polyethylene (PEX) plastic tubing as described in <u>ASTM F877,</u> ASTM F1807, ASTM F1960, and ASTM F2080 shall be installed in accordance with the manufacturer's instructions.

TABLE 605.5 Pipe Fittings

Material	Standard
Fittings for cross-linked polyethylene (PEX) plastic tubing	<u>ASTM F877;</u> ASTM F1807, ASTM F1960, ASTM F2080

*Reader's Note: *Other changes to the table have been made by the code change process and are explained in additional significant change items. Further portions of tables not covered in this book as significant changes will remain as printed from the 2003 code edition.*

Add new standard to Chapter 13 as follows:
ASTM F877-00, Standard Specification For Crosslinked Polyethylene (PEX) Plastic Hot and Cold Water Distribution Systems, Table 605.5, 605.17.2

Vanguard Crimping Ring Installation

CHANGE SIGNIFICANCE. This standard covers requirements, test methods, and methods of marking for cross-linked polyethylene plastic hot and cold water distribution systems; the components include tubing and compression fittings. This standard is currently referenced in Table 605.4 for water distribution pipe and Table 605.3 for water service pipe.

605.21

Polypropylene (PP) Plastic

Polypropylene Plastic

CHANGE TYPE. Addition

CHANGE SUMMARY. The code now contains joining methods for polypropylene plastic materials.

2006 CODE: 605.21 Polypropylene (PP) Plastic. Joints between PP plastic pipe and fittings shall comply with Sections 605.21.1 or 605.21.2.

605.21.1 Heat-Fusion Joints. Heat fusion joints for polypropylene pipe and tubing joints shall be installed with socket-type heat-fused polypropylene fittings, butt-fusion polypropylene fittings or electro fusion polypropylene fittings. Joint surfaces shall be clean and free from moisture. The joint shall be undisturbed until cool. Joints shall be made in accordance with ASTM F 2389.

605.21.2 Mechanical and Compression Sleeve Joints. Mechanical and compression sleeve joints shall be installed in accordance with the manufacturer's instructions.

CHANGE SIGNIFICANCE. The proposed text provides joining methods for polypropylene plastic pipe and fittings. The joining methods are necessary with the inclusion of polypropylene plastic pipe conforming to standard ASTM F2389 in the water service, water distribution, and pipe fittings tables.

Polypropylene Plastic, Heat Fusion

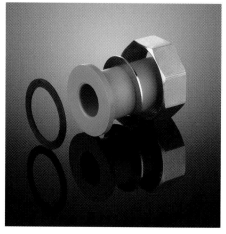

Polypropylene Plastic, Mechanical Joint

606.3

Access to Valves (Water Distribution Systems)

CHANGE TYPE. Clarification

CHANGE SUMMARY. The section was revised to clarify that all valves should be accessible.

2006 CODE: 606.3 Access to Valves. Access shall be provided to all ~~required~~ full-open valves and shutoff valves.

CHANGE SIGNIFICANCE. The current text was corrected to provide clarification. Previously it presented a situation where a valve may be installed by choice instead of being required and then not providing access. This was clearly not the intent of the code. All valves need access regardless of whether they are required by the code.

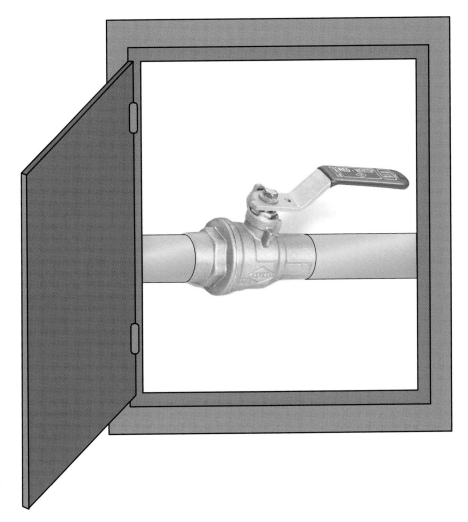

Access to Valves

607.1 and 416.5

Where Required: Hot Water and Tempered Water for Public Hand-Washing Facilities

CHANGE TYPE. Modification

CHANGE SUMMARY. The hot water requirement section was revised to reference a new tempered water safety device, ASSE 1070. A new section was added in the fixture chapter, which addresses hand-washing facility temperature protection through the use of the ASSE 1070 device.

2006 CODE: 607.1 Where Required. In residential occupancies, hot water shall be supplied to all plumbing fixtures and equipment utilized for bathing, washing, culinary purposes, cleansing, laundry, or building maintenance. In nonresidential occupancies, hot water shall be supplied for culinary purposes, cleansing, laundry, or building maintenance purposes. In nonresidential occupancies, hot water or tempered water shall be supplied for bathing and washing purposes. ~~In nonresidential occupancies, hot water or tempered water shall be supplied for bathing and washing purposes.~~ Tempered water shall be supplied through a water temperature limiting device that conforms to ASSE 1070 and shall limit the tempered water to a maximum of 110° F (43° C). This provision shall not supersede the requirement for protective shower valves in accordance with Section 424.3. ~~Tempered water shall be delivered from accessible hand-washing facilities.~~

Public Hand-Washing Plumbing Fixture

Temperature – (Tempered water) Limited to a maximum of 110° F, not less than 85° F

Controlled by - ASSE 1070-04, Performance Requirements for Water-temperature Limiting Devices

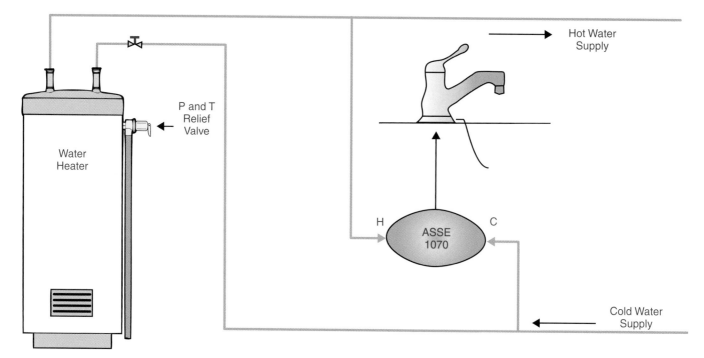

Where Required, Temperature Limiting Valves

416.5 Tempered Water for Public Hand-Washing Facilities.
Tempered water shall be delivered from public hand-washing facilities through an approved water temperature limiting device that conforms to ASSE 1070.

CHANGE SIGNIFICANCE. The revised section provides an additional level of safety to reduce the risk of scalding by adding language to clarify the application and the appropriate device to supply tempered water. Devices conforming to ASSE 1070 were recommended by the *International Plumbing Code* Ad Hoc Committee during the code cycle.

The new section 416.5 information was originally contained in 607.1. It has been relocated to section 416, which addresses lavatories in the fixture table. Protecting public hand-washing facilities with this device will reduce the risk of scalding.

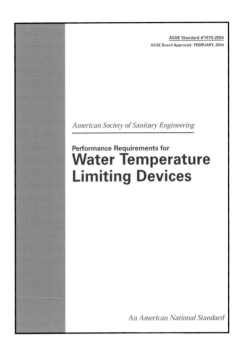

ASSE Standard #1070-2004
ASSE Board Approved: FEBRUARY, 2004

American Society of Sanitary Engineering

Performance Requirements for
**Water Temperature
Limiting Devices**

An American National Standard

Where Required, Temperature Limiting Valves

607.4

Flow of Hot Water to Fixtures

Flow of Hot Water to Fixtures

CHANGE TYPE. Clarification

CHANGE SUMMARY. The section with its exception was revised to include Canadian Standards Association (CSA) standard devices and to clarify the section's intent.

2006 CODE: 607.4 Flow of Hot Water to Fixtures. Fixture fittings, ~~and~~ faucets, <u>and diverters</u> ~~that are supplied with both hot and cold water~~, shall be installed and adjusted so that <u>the flow of hot water from the fittings corresponds</u> to the left-hand side of the ~~water temperature control~~ <u>fixture fitting</u> ~~represents the flow of hot water, when facing the outlet~~.

Exception:
Shower and tub/shower mixing valves conforming to ASSE 1016 or <u>CSA B125</u>, where the <u>flow of hot water</u> ~~temperature control~~ corresponds to the markings on the device.

CHANGE SIGNIFICANCE. The revision recognizes the CSA B125 standards device in the exception, which has been previously approved and referenced in other locations of the code. The change improves the main section and its exception in identifying the control rather than focusing on the valve's interior piping. Further, the reference to hot and cold water supply indicates that tempered water shower devices are not considered in this section.

CHANGE TYPE.   Modification

CHANGE SUMMARY.   The backflow prevention application table has been revised to include additional CSA standards.

Table 608.1

Application of Backflow Preventers

TABLE 608.1   **Application of Backflow Preventers**

Device	Degree of Hazard	Application B	Applicable Standards
Reduced pressure principle backflow preventer and reduced pressure principle fire protection backflow preventer	High or low hazard	Backpressure or back-siphonage Sizes $3/8''$–16″	~~CAN/~~CSA B64.4, CSA B64.4.1
Double check backflow prevention assembly and double check fire protection backflow prevention assembly	Low hazard	Backpressure or back-siphonage Sizes $3/8''$–16″	CSA B64.5, CSA B64.5.1
Dual-check-valve-type backflow preventer	Low hazard	Backpressure or back-siphonage Sizes $1/4''$–1″	CSA B64.6, CSA B64.6.1
Backflow preventer with intermediate atmospheric vents	Low hazard	Backpressure or back-siphonage Sizes $1/4''$–$3/4''$	~~CAN/~~CSA B64.3
Backflow preventer for carbonated beverage machines	Low hazard	Backpressure or back-siphonage Sizes $1/4''$–$3/8''$	CSA B64.3.1
Pipe-applied atmospheric-type vacuum breaker	High or low hazard	Backsiphonage only Sizes $1/4''$–4″	~~CAN/~~CSA B64.1.1
Pressure vacuum breaker assembly	High or low hazard	Backsiphonage only Sizes $1/2''$–2″	CSA B64.1.2
Hose-connection vacuum breaker	High or low hazard	Low head backpressure or backsiphonage Sizes $1/2,'' 3/4,''$ 1″	~~CAN/~~CSA B64.2, CSA B64.2.1
Vacuum breaker wall hydrants, frost-resistant, automatic draining type	High or low hazard	Low head backpressure or backsiphonage Sizes $3/4''$–1″	~~CAN/~~CSA B64.2.2
Hose connection backflow preventer	High or low hazard	Low head backpressure, rated working pressure backpressure or back-siphonage Sizes $1/2''$–1″	CSA B64.2.1.1

*Reader's Note: *Portions of table not covered in this book as signifi-cant changes will remain as printed from the 2003 code edition.*

Table 608.1 continues

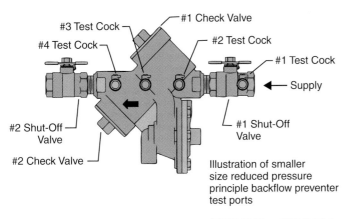

#3 Test Cock
#4 Test Cock
#1 Check Valve
#2 Test Cock
#1 Test Cock
Supply
#2 Shut-Off Valve
#2 Check Valve
#1 Shut-Off Valve

Illustration of smaller size reduced pressure principle backflow preventer test ports

ASSE 1013 or CSA B 64.4

Application of Backflow Preventers

Table 608.1 continued

CHANGE SIGNIFICANCE. The additional CSA devices referenced by the approved standards will enable manufacturers with products certified to the applicable CSA standard to have their products used as options to products that meet the requirements of the current ASSE standards. This change will also allow the authorities having jurisdiction to allow the use of products that meet either CSA or ASSE Standards.

The new standards recognized in the table above were approved in separate code-approval actions. The standards are listed below for the reader's convenience, with the appropriate section number, title, and change action reference.

1. CSA B64.4.1, 608.13.2, Reduced pressure principle backflow presenters, Approved As Submitted (Proposal P73-03/04)

2. CSA B64.1.2, 608.13.5, Pressure-type vacuum breakers, Approved As Submitted (Proposal P74-03/04)

3. CSA B64.2.1, CSA B64.2.1.1, 608.13.6, Atmospheric-type vacuum breakers, Approved As Submitted (Proposal P75-03/04)

4. CSA B64.5, CSA B64.5.1, 608.13.7, Double check-valve assemblies, Approved As Submitted (Proposal P76-03/04)

5. CSA B64.3.1, 608.16.1, Beverage dispensers, Approved As Submitted (Proposal P78-03/04)

• Large reduced pressure principle back flow preventer (RPZ) horizontal application ASSE 1013 or CSAB 64.4

• The best method of back flow protection is an air gap (separation), the next choice is an RPZ

Supply

Application of Backflow Preventers

CHANGE TYPE.   Clarification

CHANGE SUMMARY.   The section now includes a specific conformance standard for backflow prevention in fixture fittings.

2006 CODE: 608.2 Plumbing Fixtures.   The supply lines and fittings for every plumbing fixture shall be installed as to prevent backflow. <u>Plumbing fixture fittings shall provide backflow protection in accordance with ASME A112.18.1.</u>

Chapter 13 Referenced Standards
ASME A112.18.1–2000 Plumbing Fixture Fittings . . . 424.1, <u>608.2.</u>

CHANGE SIGNIFICANCE.   The revision adds ASME A112.18.1, a standard specifically developed for backflow prevention devices for use in plumbing fixture fittings (see Section 424.1). This standard also includes individual backflow prevention devices covered in Table 608.1.

608.2

Plumbing Fixtures (Protection of the Potable Water Supply)

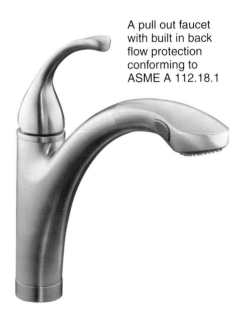

A pull out faucet with built in back flow protection conforming to ASME A 112.18.1

Plumbing Fixtures
(Protection of the Potable Water Supply)

608.16.10

Coffee Machines and Noncarbonated Beverage Dispensers

Coffee Machines and
Noncarbonated Beverage Dispensers

CHANGE TYPE.   Addition

CHANGE SUMMARY.   A section has been added to identify that backflow protection is required for coffee machines and noncarbonated beverage dispensers.

2006 CODE: <u>**608.16.10 Coffee Machines and Noncarbonated Beverage Dispensers.**   The water supply connection to coffee machines and noncarbonated beverage dispensers shall be protected against backflow by a backflow preventer conforming to ASSE 1022 or by an air gap.</u>

CHANGE SIGNIFICANCE.   Coffee machines and noncarbonated beverage dispensers have the potential for backsiphonage and should be protected by a backflow prevention device or by an air gap integral with the equipment. An ASSE 1022 Backflow Preventer for Beverage Dispensing Equipment is presently required for beverage dispensing equipment.

The ASSE 1022 device includes two check valves and an atmospheric vent. This device is somewhat similar to the protection accepted on a boiler without chemical additives. Many code officials for several years have considered espresso machines as enclosed systems similar to mini boilers.

CHANGE TYPE.   Addition

CHANGE SUMMARY. The Underground drainage and vent table were revised to include the ASTM F1412 standard, and the pipe fittings table was revised to include polyolefin fittings conforming to ASTM F1412 or CSA B181.3 standards. The later change referenced above added joining criteria for the polyolefin products.

TABLE 702.2   Underground Building Drainage and Vent Pipe

Material	Standard
Polyolefin pipe	ASTM F1412

TABLE 702.4   Pipe Fittings

Material	Standard
Polyolefin	ASTM F1412, CSA B181.3

Reader's Note: *Portions of tables not covered in this book as significant changes will remain as printed from the 2003 code edition.*

Chapter 13 Referenced Standards
CSA B181.3-02 Polyolefin Laboratory Drainage Systems . . . . Table 702.4
ASTM F1412-01 Specification for Polyolefin Pipe and Fittings for Corrosive Waste Drainage . . . . Table 702.2, Table 702.4

2006 CODE: 705.17 Polyolefin Plastic. Joints between polyolefin plastic pipe and fittings shall comply with Sections 705.17.1 and 705.17.2.

705.17.1 Heat-Fusion Joints. Heat fusion joints for polyolefin pipe and tubing joints shall be installed with socket-type heat-fused polyolefin fittings or electrofusion polyolefin fittings. Joint surfaces shall be clean and free from moisture. The joint shall be undisturbed until cool. Joints shall be made in accordance with ASTM F 1412 or CSA B181.3.

705.17.2 Mechanical and Compression Sleeve Joints. Mechanical and compression sleeve joints shall be installed in accordance with the manufacturer's instructions.
(Renumber remaining sections)

CHANGE SIGNIFICANCE. The revision to the underground drainage and vent table now includes the ASTM F1412 standard plus the existing CSA B181.3 standard for polyolefin products. The revision to the drain, waste, and vent piping and fitting materials table now includes both the ASTM F1412 and CSA B181.3 standards. The product was recognized but the standards were absent in both tables. This will al-

Table 702.2, Table 702.4, and 705.17 continues

Table 702.2, Table 702.4, and 705.17

Underground Building Drainage and Vent Pipe, Pipe Fittings, and Polyolefin Plastic (Joints)

Polyolefin Plastic Fittings

Polyolefin Plastic Fittings,
Heat-Fusion Joints

Table 702.2, Table 702.4, and 705.17 continued

low the use of products made in conformance with the ASTM standards as well as those made to the CSA standard and will eliminate the need to certify products to both standards. ASTM F1412 was first adopted in 1992 and has been used successfully for more than 10 years with state and local approval.

The later change referenced above added joining criteria for polyolefin products to the joining methods referenced in Section 705, Joints. Sections 705.1 to 705.20 are not in alphabetical order; however, they address methods for ABS plastic to vitrified clay material joints.

705.5.2 and 705.5.3

Compression Gasket Joints and Mechanical Joint Coupling

CHANGE TYPE.   Modification

CHANGE SUMMARY.   Two cast iron joining method sections were revised to include new standards for compression gasket joints in hub and spigot and hubless pipe and fittings.

2006 CODE: 705.5.2 Compression Gasket Joints.   Compression gaskets for hub and spigot pipe and fittings shall conform to ASTM C 564 and shall be tested to ASTM C1563. Gaskets shall be compressed when the pipe is fully inserted.

705.5.3 Mechanical Joint Coupling.   Mechanical joint couplings for hubless pipe and fittings shall comply with CISPI 310, ~~or~~ ASTM C 1277, or ASTM C1540. The elastomeric sealing sleeve shall conform to ASTM C564 or CAN/CSA B602 and shall be provided with a center stop. Mechanical joint couplings shall be installed in accordance with the manufacturer's installation instructions.

705.5.2 and 705.5.3 continues

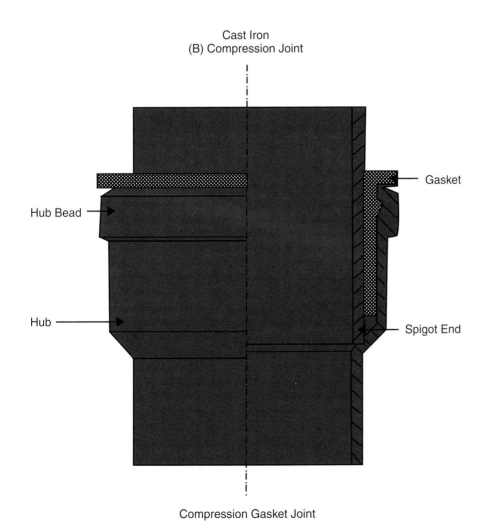

Cast Iron
(B) Compression Joint

Gasket

Hub Bead

Hub

Spigot End

Compression Gasket Joint

705.5.2 and 705.5.3 continued

CHANGE SIGNIFICANCE. The section revision for compression gaskets for hub and spigot pipe and fittings recognizes a new ASTM standard. C1563 establishes test criteria for the finished gaskets, which include hydrostatic tests of the finished gaskets in typical field installations. The gaskets themselves are required to be manufactured to ASTM material standard ASTM C564.

The section revision for mechanical joint couplings for hubless pipe and fittings recognizes the new standard ASTM C1540, Standard Specifications for Heavy Duty Shielded Couplings Joining Hubless Cast Iron Soil Pipe and Fittings. The elastomeric sealing sleeve shall conform to material standards ASTM C564 or CSA B602 and shall have a center stop.

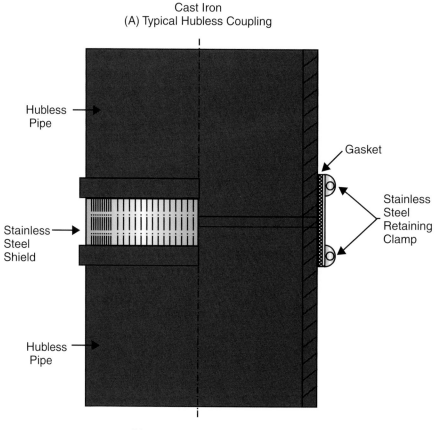

Cast Iron
(A) Typical Hubless Coupling

Hubless Pipe

Gasket

Stainless Steel Shield

Stainless Steel Retaining Clamp

Hubless Pipe

Mechanical Joint Coupling

CHANGE TYPE. Addition

CHANGE SUMMARY. Sections were added to identify how polyethylene plastic pipe and fittings can be made by heat fusion and mechanical means. Further, the tools required by manufacturer's instructions have been added.

2006 CODE: 705.16 Polyethylene Plastic Pipe. Joints between polyethylene plastic pipe and fittings shall be underground and shall comply with Sections 705.16.1 or 705.16.2.

705.16.1 Heat Fusion Joints. Joint surfaces shall be clean and free from moisture. All joint surfaces shall be cut, heated to melting temperature, and joined using tools specifically designed for the operation. Joints shall be undisturbed until cool. Joints shall be made in accordance with ASTM D 2657 and the manufacturer's instructions.

705.16.2 Mechanical Joints. Mechanical joints in drainage piping shall be made with an elastomeric seal conforming to ASTM C 1173, ASTM D 3212, or CAN/CSA B602. Mechanical joints shall be installed in accordance with the manufacturer's instructions.

CHANGE SIGNIFICANCE. Polyethylene piping material was approved for use in Table 702.3, Building Sewer Pipe, of the *International Plumbing Code* 2003 edition. However, joining methods were not addressed. Joining methods for polyethylene were added in the 03/04 code update cycle referencing heat fusion and mechanical joints.

The revision proposal during the 04/05 code hearings stated that "this material was added into the code based on the reason it would allow trenchless installations of sewers." The proponent requested a clarification be added to Section 705.16 requiring polyethylene only be installed underground. The underground restriction is based upon the material only appearing in the building sewer pipe table, 702.3.

705.16
Polyethylene Plastic Pipe

706.4

Heel- or Side-Inlet Quarter Bends

CHANGE TYPE. Addition

CHANGE SUMMARY. The section was added to identify and establish use criteria for heel-inlet quarter bends and side-inlet quarter bends.

2006 CODE: <u>**706.4 Heel- or Side-Inlet Quarter Bends.** Heel-inlet quarter bends shall be an acceptable means of connection, except where the quarter bends serves a water closet. A low-heel inlet shall not be used as a wet-vented connection. Side-inlet quarter bends shall be an acceptable means of connection for drainage, wet venting, and stack venting arrangements.</u>

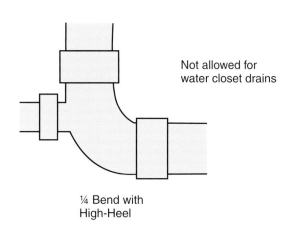

Not allowed for
water closet drains

¼ Bend with
High-Heel

A. Not allowed for
water closet drains

and

B. Low-heel inlet not
allowed for connection
of a wet vented fixture drain

¼ Bend with
Low-Heel

Heal Inlet Quarter Bend

CHANGE SIGNIFICANCE. The added section includes provisions for the use of heel-inlet quarter bends and side-inlet quarter bends in Section 706, Connections Between Drainage Piping and Fittings. These fittings are presently referenced and allowed in the *International Residential Code.*

706.4 continues

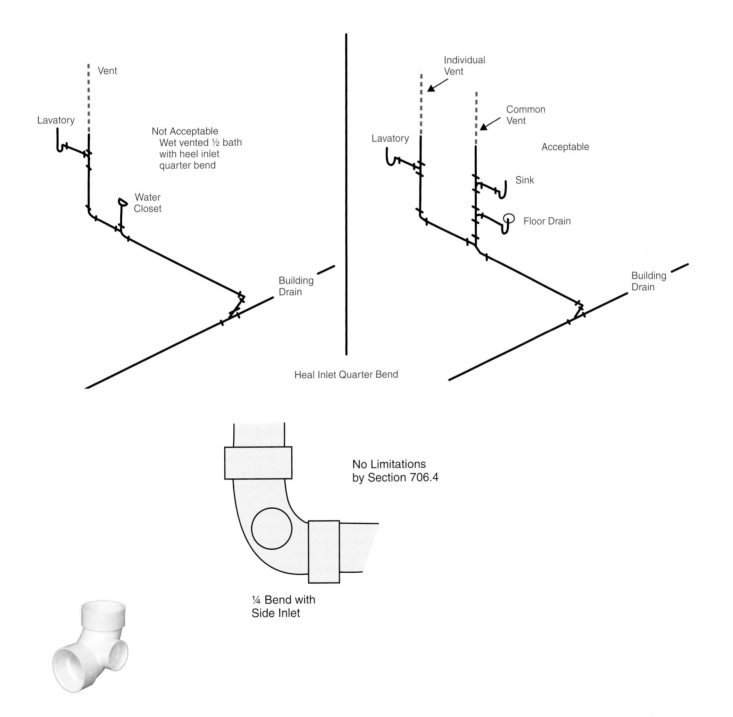

Heal Inlet Quarter Bend

Side Inlet Quarter Bend

706.4 continued

These fittings have been in existence for many years and are illustrated in installation graphics for many different codes. This added code text considers the flow volume in the fitting pattern and eliminates water closet discharges, which would seriously affect other fixtures connected with the heel inlet.

Code users often notice that drainage waste and vent fittings constructed of various materials have different names for the fitting patterns. Also, manufacturers of other material fittings may have different terms for fittings with different lay lengths (for example, 90° quarter bend long sweeps).

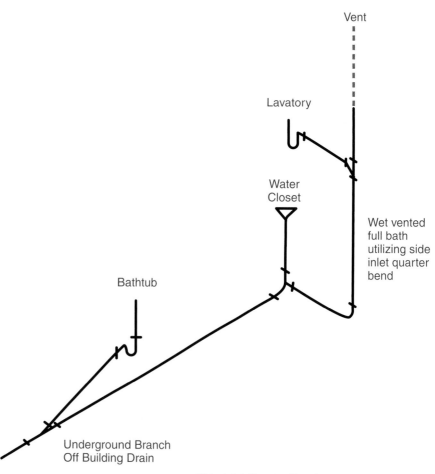

Side Inlet Quarter Bend

708.3.3
Changes of Direction (for Cleanouts)

CHANGE TYPE. Modification

CHANGE SUMMARY. This change requires building sewers to have the same cleanout installation conditions as the building's drainage system piping.

2006 CODE: 708.3.3 Changes of Direction. Cleanouts shall be installed at each change of direction <u>greater than 45 degrees (0.79 rad)</u> in ~~of~~ the <u>building sewer,</u> building drain, <u>and</u> ~~or~~ horizontal waste or soil lines. ~~greater than 45 degrees (0.79 rad).~~ Where more than one change of direction occurs in a run of piping, only one cleanout shall be required for each 40 feet (12 192 mm) of developed length of the drainage piping.

CHANGE SIGNIFICANCE. This revision requires cleanouts for all building sewers to be installed the same as building drains and horizontal waste lines that are in the buildings. Designers and code officials have usually applied cleanout requirements for building sewers when the directional change was greater than 45 degrees. However, the cleanout section specific to building sewers, Section 708.3.2, did not mandate a cleanout for changes greater than 45 degrees, except for sewers greater than 8 inches.

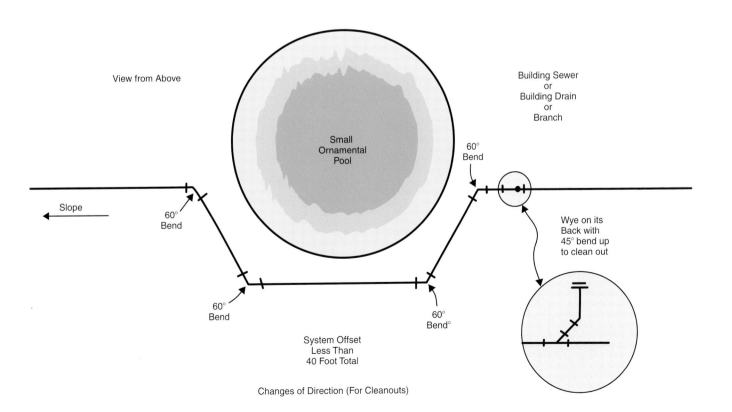

Changes of Direction (For Cleanouts)

Table 709.1

Drainage Fixture Units for Various Fixture Groups

CHANGE TYPE. Modification

CHANGE SUMMARY. The table was revised to include an entry for service sinks. A revision also addressed urinals not having water supplied for flushing purposes.

TABLE 709.1 Drainage Fixture Units for Fixtures and Groups

Fixture Type	Drainage Fixture Unit Value as Load Factors	Minimum Size of Trap (Inches)
Service sink	2	1½
Urinal, non-water supplied	0.5	Note d

*Reader's Note: *Portions of tables not covered in this book as significant changes will remain as printed from the 2003 code edition.*

CHANGE SIGNIFICANCE. The first approved code change now lists service sinks as a new entry. The 1995 edition of the *International Plumbing Code* listed service sinks under the heading of sinks, which was later revised to include sinks in general. This code change will improve understanding of the commonly installed fixture.

The second code change assigns a reduced drainage fixture value for urinals not having conventional water supplies. Section 419.1, which addresses urinal plumbing fixtures, was revised in proposals P27-04/05 and P28-03/04 to include language for these urinals and standards conformance to ANSI Z124.9.

Drainage Fixture Units for Fixture Groups

The proponent provided the following reduction justification. A nonflushing or waterless urinal has a significantly lower contribution to the drainage system flow than a flush urinal. A standard urinal flushes 1 gallon of water with each flush. The average quantity of urine excreted when using a urinal is between 1 and $1\frac{1}{2}$ pints. Thus, the total discharge into the drainage system for a standard flush urinal is less than 1.2 gallons of waste. The minimum interval between use is 43 seconds. For a waterless urinal, the interval between use remains the same. However, the discharge into the drainage systems is less than 0.2 gallons. This equates to a dfu value of less than 0.5. However, the value is rounded up to 0.5 for convenience in sizing the drainage system. As with flush urinals, the trap size is determined by the fixture manufacturer.

712.2

Valves Required (for Sumps and Ejectors)

CHANGE TYPE.   Modification

CHANGE SUMMARY.   The section was revised to delete the exception, which differed from information addressing one- and two-family dwellings in the *International Residential Code.*

2006 CODE: 712.2 Valves Required.   A check valve and full open valve located on the discharge side of the check valve shall be installed in the pump or ejector discharge piping between the pump or ejector and the gravity drainage system. Access shall be provided to such valves. Such valves will be located above the sump cover required by Section 712.1 or, where the discharge pipe from the ejector is below grade, the valves shall be accessibly located outside the sump below grade in an access pit with a removable access cover.

~~Exception:~~

~~In one- and two-family dwellings, only a check valve shall be required, located on the discharge piping from the sewage pump or ejector.~~

Valves Required (For Sumps and Ejectors)

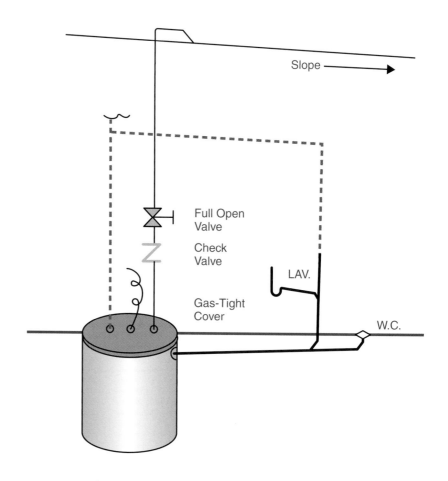

Valves Required (For Sumps and Ejectors)

CHANGE SIGNIFICANCE. The text that is shown to be deleted was in conflict with Section P3007.1 of the *International Residential Code,* which requires a check valve and a gate valve installed on the discharge piping between the pump and ejector and the drainage system. In addition, this text applies to one- and two-family dwellings, and Section 101.2, exception 1, of the *International Plumbing Code* states that "detached one and two family dwellings must comply with the *International Residential Code* (IRC)."

Originally the residential code contained the exception language. However, subsequent code changes updated the pump servicing requirement to be the same as the *International Plumbing Code* (IPC). The discussion here serves as a reminder that many enforcement jurisdictions adopt the IPC only. There are minor differences in the plumbing area between the codes, by design. It has been understood that the residential applications have referenced greater code flexibility because of occupant loading and frequency-of-use considerations.

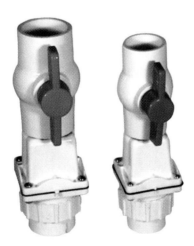

30-0100
1½" PVC Check Valve / Ball Valve / Union Combination
30-0101
2" PVC Check Valve / Ball Valve / Union Combination

Examples of Combination Products

903

Outdoor Vent Extension

CHANGE TYPE. Clarification

CHANGE SUMMARY. The titles Vent Stacks and Stack Vents, Sections 903.1 and 903.1.1, were deleted. New sections 903.1, 903.1.1, and 903.1.2 were added with the changed section heading to address Outdoor Vent Extensions.

2006 CODE: ~~Vent Stacks and Stack Vents~~

~~**903.1 Stack Required.** Every building in which plumbing is installed shall have at least one stack the size of which is not less than one half of the required size of the building drain. Such stack shall run undiminished in size and as directly as possible from the building drain through to the open air or to a vent header that extends to the open air.~~

~~**903.1.1 Connection to drainage system.** A vent stack shall connect to the building drain or to the base of a drainage stack in accordance with Section 903.4. A stack vent shall be an extension of the drainage stack.~~

Outdoor Vent Extension

903.1 Required vent extension. The vent system serving each building drain shall have at least one vent pipe that extends to the outdoors.

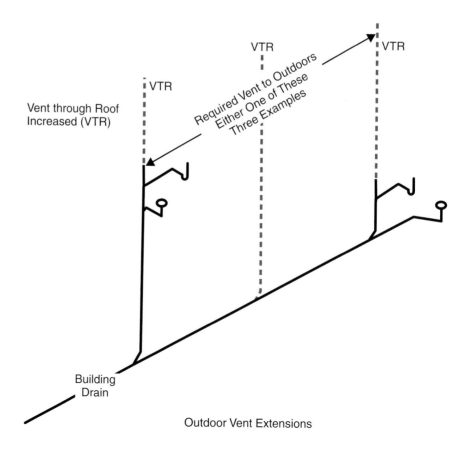

Outdoor Vent Extensions

903.1.1 Installation. The required vent shall be a dry vent that connects to the building drain or an extension of a drain that connects to the building drain. Such vent shall not be an island fixture vent as allowed by Section 913.

903.1.2 Size. The required vent shall be sized in accordance with Section 916.2 based on the required size of the building drain.

CHANGE SIGNIFICANCE. The intent of the code is to ensure that the vent systems contain a pipe extension that will eventually end up outdoors. The code change proponent pointed out several areas of confusion in the past two sections and section heading, which were removed. Previous Section 903.1 referenced stack vents that might not have been provided for structures with a slab on grade with vents connecting only to fixture branches. The proponent was concerned about how a stack could be provided and whether a stack vent could be provided. Also, the term *undiminished* was often misapplied to mean undiminished size of the building drain through the vent, which is incorrect. The code gives provisions for sizing venting systems in other sections of the venting chapter on sizing compliance. Finally, it is a common practice to offset a vent out the back and through the roof to avoid visible penetrations of the front portion of a building. This common practice of an offset was misunderstood as not being as "directly as possible."

Previous section 903.1.1 only contained relevant information for the location in which a vent stack is permitted to connect. That information is presently in current Section 903.4. Further, several stack vents may extend into a branch vent and have only one extension to the outdoors, because not all stack vents have to extend separately to the outdoors.

The new section 903.1 clearly identifies the intent that at least one vent shall terminate outdoors. This requirement guarantees that positive pressures in the system will be relieved.

Section 903.1.1 ensures a dry vent will provide sufficient air circulation to the building drain for correct performance and identifies island fixture vents as not allowed. Island fixture vents shall not be considered for all fixtures, as clarified in section 913.1.

The third new section, 903.1.2, provides vent sizing for the one required vent based upon the building drain and 916.2 of the code.

903.3 and 917.1 through 917.4

Vent Termination and Air Admittance Valves

CHANGE TYPE. Modification

CHANGE SUMMARY. Section 903.3 is expanded to include stack type air admittance valves and the sections in 917 are revised to clarify criteria for stack and branch type air admittance valves.

2006 CODE: 903.3 Vent Termination. Every vent stack or stack vent shall ~~extend outdoors and~~ terminate outdoors to the open air or to a stack type air admittance valve in accordance with Section 917.

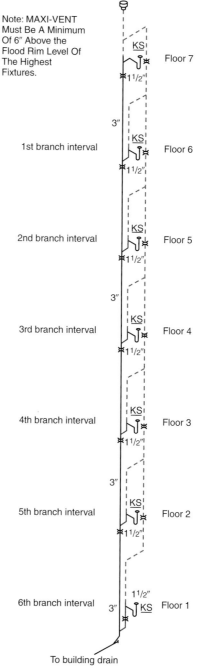

Note: MAXI-VENT Must Be A Minimum Of 6″ Above the Flood Rim Level Of The Highest Fixtures.

1st branch interval

2nd branch interval

3rd branch interval

4th branch interval

5th branch interval

6th branch interval

To building drain

Vent Termination and Air Admittance Valves

917.1 General. Vent systems utilizing air admittance valves shall comply with this section. <u>Stack type air admittance valves shall conform to ASSE 1050.</u> Individual and branch type air admittance valves shall conform to ASSE 1051.

917.2 Installation. The valves shall be installed in accordance with the requirements of this section and the manufacturer's installation instructions. Air admittance valves shall be installed after the DWV testing required by Section 312.2 or 312.3 has been performed.
   (Existing text)

917.3 Where Permitted. Individual, branch, and circuit vents shall be permitted to terminate with a connection to an <u>individual or branch type</u> air admittance valve. <u>Stack vents and vent stacks shall be permitted to terminate to stack type air admittance valves.</u> ~~The~~ <u>Individual and branch type</u> air admittance valves shall only vent fixtures that are on the same floor level and connect to a horizontal branch drain. The horizontal branch drain <u>having individual and branch type air admittance valves</u> shall conform to Section 917.3.1 or 917.3.2. <u>Stack type air admittance valves shall conform to Section 917.3.3.</u>

903.3 and 917.1 through 917.4 continues

OPEN

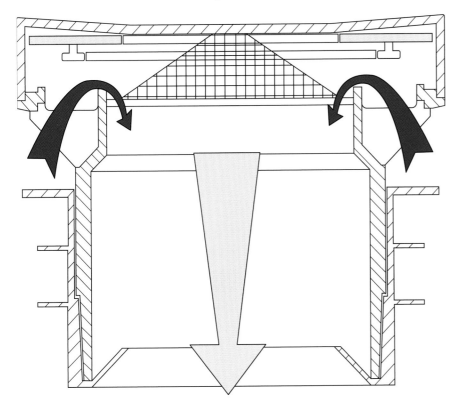

Air admittance valve opens to
admit air to relieve negative pressure

903.3 and 917.1 through 917.4
continued

917.3.1 Location of Branch. The horizontal branch drain shall connect to the drainage stack building drain a maximum of four branch intervals from the top of the stack.

(Existing text)

917.3.2 Relief Vent. <u>Where the horizontal branch is located more than four branch intervals from the top of the stack,</u> ~~T~~<u>the</u> horizontal branch shall be provided with a relief vent that shall connect to a vent stack or stack vent, or extend outdoors to the open air. The relief vent shall connect to the horizontal branch drain between the stack ~~or building drain~~ and the most downstream fixture drain connected to the horizontal branch drain. The relief vent shall be sized in accordance with Section 916.2 and installed in accordance with Section 905. The relief vent shall be permitted to serve as the vent for other fixtures.

917.3.3 Stack. <u>Stack type air admittance valves shall not serve as the vent terminal for vent stacks or stack vents that serve drainage stacks exceeding six branch intervals.</u>

917.4 Location. ~~The~~ <u>Individual and branch type</u> air admittance valves shall be located a minimum of 4 inches (102 mm) above the horizontal branch drain or fixture drain being vented. <u>Stack type air admittance valves shall be located not less than 6 inches (152 mm)</u>

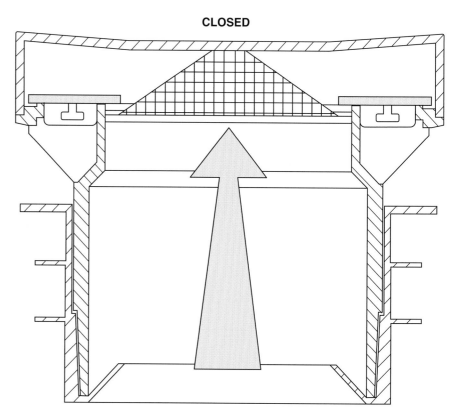

CLOSED

Air admittance valve closes and
seals under zero or positive pressures

<u>above the flood level rim of the highest fixture being vented</u>. The air admittance valve shall be located within the maximum developed length permitted for the vent. The air admittance valve shall be installed a minimum of 6 inches (152 mm) above insulation materials.

CHANGE SIGNIFICANCE.   These code changes allow the use of air admittance valves that conform to standard ASSE 1050 for stack applications. The valves operate the same as previously approved branch installation valves conforming to ASSE 1051. They open under negative pressure, allowing air to enter the system, and close under neutral or positive pressure conditions, preventing sewer gases from escaping from the valve. The one open pipe vent required on every building drainage system in Section 917.7, Vent Required, and Section 903.1, Required Vent Extension, mandate that at least one vent pipe shall extend to the outdoors to relieve the system's positive pressure.

903.3 and 917.1 through 917.4 continues

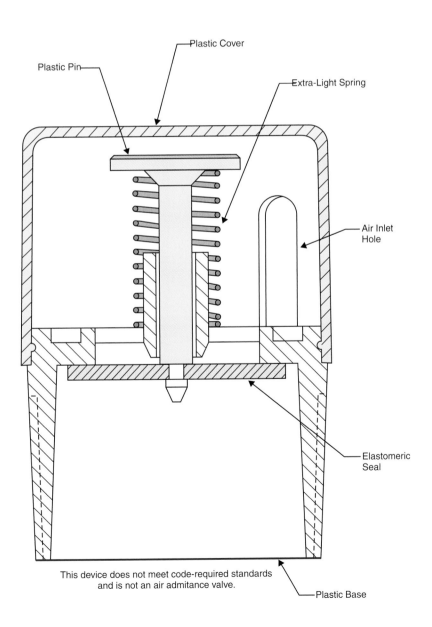

Plastic Cover

Plastic Pin

Extra-Light Spring

Air Inlet Hole

Elastomeric Seal

This device does not meet code-required standards and is not an air admitance valve.

Plastic Base

*903.3 and 917.1 through 917.4
continued*

Since the code revisions identify stack devices, several sections must be clarified for different applications between existing recognized branch devices, ASSE 1051, and newly recognized stack type devices, ASSE 1050.

Two significant branch interval considerations are important to note in the changes. First, the stack type devices are limited to stacks having branch intervals of 6 or less, in Section 917.3.3. Second, Section 917.3.2 in its first sentence again points out that not every branch utilizing a branch air admittance valve needs a relief vent. Section 917.3 states that horizontal branch drains shall conform to 917.3.1 "or" 917.3.2. Relief vents are required only for branches that connect to the drainage stack below the top four branch intervals from the top of the stack.

906.1

Distance of Trap from Vent (Water Closet from Vent)

CHANGE TYPE. Modification

CHANGE SUMMARY. The section has been revised to include an exception that does not limit the distance of a water closet from its required vent.

2006 CODE: 906.1 Distance of Trap from Vent. Each fixture trap shall have a protecting vent located so that the slope and the developed length in the fixture drain from the trap weir to the vent fitting are within the requirements set forth in Table 906.1.

> **Exception:**
> The developed length of the fixture drain from the trap weir to the vent fitting for self-siphoning fixtures, such as water closets, shall not be limited.

CHANGE SIGNIFICANCE. This change is intended to clarify the intent of the code and does not change any requirements for the fixture drain of self-siphonage fixtures other than the trap-to-vent distance. A vent is still required, but the distance to the vent is not limited. The approved text is identical to a section located in the *International Residential Code,* Section P3105.1.

906.1 continues

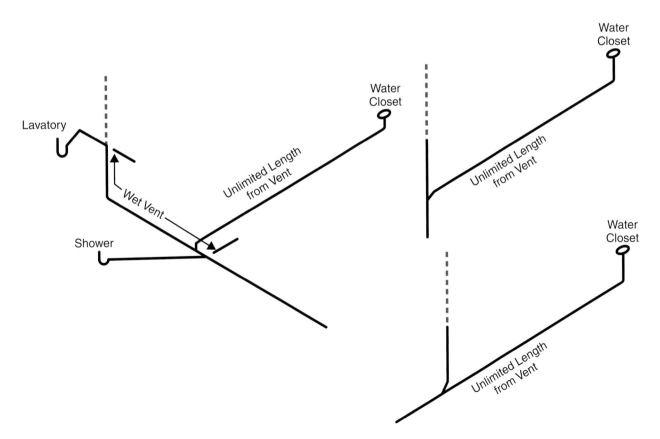

Distance of Trap from Vent (For Water Closets)

906.1 continued

The distance from trap to vent is intended to protect the trap seal against self-siphonage. A water closet and siphon-action urinal do not have to be protected against self-siphonage since the fixtures must siphon to properly operate. These fixtures are also unique in that they reseal the trap after each use. The code is often misinterpreted as requiring a trap-to-vent distance for water closets and siphon action urinals.

CHANGE TYPE. Modification

CHANGE SUMMARY. The table is revised to provide distance corrections consistent with engineering principals and now reads the same as the venting table in the *International Residential Code.* Further, the section was revised to clarify the trap-to-vent distances when drain lines might be increased or when the slope is increased.

Table 906.1 and Section 906.2

Maximum Distance of Fixture Trap from Vent and Venting of Fixture Drains

2006 CODE:

TABLE 906.1   Maximum Distance of Fixture Trap from Vent

Size of Trap (Inches)	Size of Fixture Drain (Inches)	Slope (Inch Per Foot)	Distance from Trap (Feet)
~~1¼~~	~~1¼~~	~~¼~~	~~3½~~
1¼	~~1½~~	¼	5
~~1½~~	~~1½~~	~~¼~~	~~5~~
1½	~~2~~	¼	6
2	~~2~~	¼	~~6~~ 8
3	~~3~~	⅛	~~10~~ 12
4	~~4~~	⅛	~~12~~ 16

Table 906.1 and Section 906.2 continues

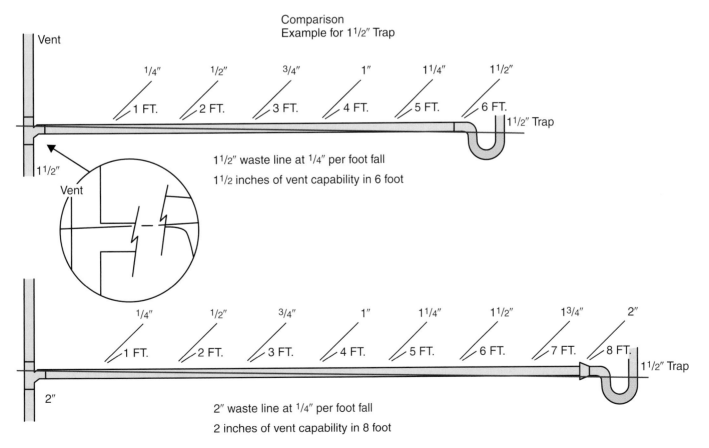

Maximum Distance of Trap from Vent

Table 906.1 and Section 906.2 continued

906.2 Venting of Fixture Drains.   ~~The vent for a fixture drain, except where serving a fixture with integral traps, such as water closets, shall connect above the weir of the fixture trap being vented.~~ <u>The total fall in a fixture drain due to pipe slope shall not exceed the diameter of the fixture drain, nor shall the vent connection to a fixture drain, except for water closets, be below the weir of the trap.</u>

CHANGE SIGNIFICANCE.   The table sets the bottom line, so to speak, for distance limits of a fixture trap from its vent. The entire column of information for "size of fixture drain" is now removed and addressed by Section 906.2.

Table 906.1 is based on the concept of applying adequate amounts of air to a fixture trap for proper operation. The basic concept proven to prevent trap seal loss is now found in the table. The total fall must not exceed one pipe diameter, or an "S" trap effect is potentially created. The most basic information is now contained in the table, which was taken from the *International Residential Code*. For example, a 1¼-inch drain line can extend 5 feet and should not be limited to the former 3½ feet (e.g., $5 \times \frac{1}{4}'' = 1\frac{1}{4}''$), and a 2-inch drain is entitled to an 8-foot rather than 6-foot allowance (e.g., $8 \times \frac{1}{4}'' = 2''$).

The Section 906.2 revision now explains and sets the limits on all the installation variations. Those variations could include selecting a drain line size larger than the trap, trying to gain more distance, or installing a drain line with greater slope because of structural limitations that reduce the allowed travel distance.

The language that exempts water closets is relocated text. This concept recognized that water closets are self-siphoning fixtures, as pointed out by the code change exception in 906.1. A vent is still required for the water closet; however, the distances may be extended since the fixture trap is replenished during the fill cycle.

CHANGE TYPE. Clarification

CHANGE SUMMARY. The three separate revisions specifically recognize horizontal and vertical wet vent concepts with their separate individual requirements.

2006 CODE: 909.1 Horizontal Wet Vent Permitted. Any combination of fixtures within two bathroom groups located on the same floor level is permitted to be vented by a horizontal wet vent. The wet vent shall be considered the vent for the fixtures and shall extend from the connection of the dry vent along the direction of the flow in the drain pipe to the most downstream fixture drain connection to the horizontal branch drain. Only the fixtures within the bathroom groups shall connect to the wet-vented horizontal branch drain. Any additional fixtures shall discharge downstream of the horizontal wet vent.

909.1.1 Vertical Wet Vent Permitted. Any combination of fixtures within two bathroom groups located on the same floor level is permitted to be vented by a vertical wet vent. The vertical wet vent shall be considered the vent for the fixtures and shall extend from the

909.1, 909.1.1, 909.2, and 909.3 continues

909.1, 909.1.1, 909.2, and 909.3

Horizontal Wet Vent Permitted, Vertical Wet Vent Permitted, Vent Connection and Size

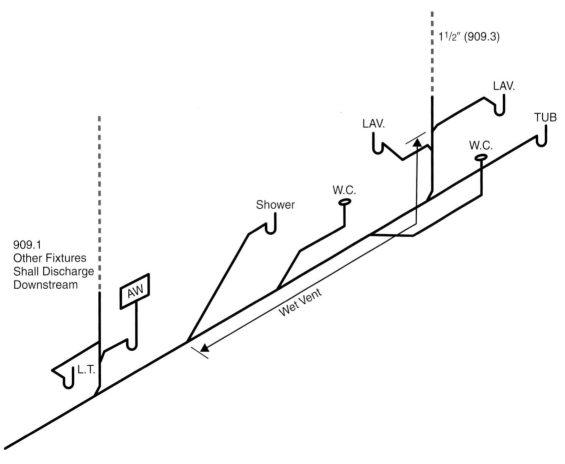

Horizontal Wet Vent

909.1, 909.1.1, 909.2, and 909.3
continued

connection of the dry vent down to the lowest fixture drain connection. Each <u>wet-vented</u> fixture shall connect independently to the vertical wet vent. Water closet drains shall connect at the same elevation. Other fixture drains shall connect above or at the same elevation as the water closet fixture drains. The dry vent connection to the vertical wet vent shall be an individual or common vent serving one or two fixtures.

909.2 Vent Connection. The dry-vent connection to the wet vent shall be an individual vent or common vent to the lavatory, bidet, shower, or bathtub. ~~The dry vent shall be sized based on the largest required diameter of pipe within the wet vent system served by the dry vent.~~ <u>In vertical wet-vent systems, the most upstream fixture drain connection shall be a dry-vented fixture drain connection. In horizontal wet vent systems, not more than one wet-vented fixture drain shall discharge upstream of the dry-vented fixture drain connection.</u>

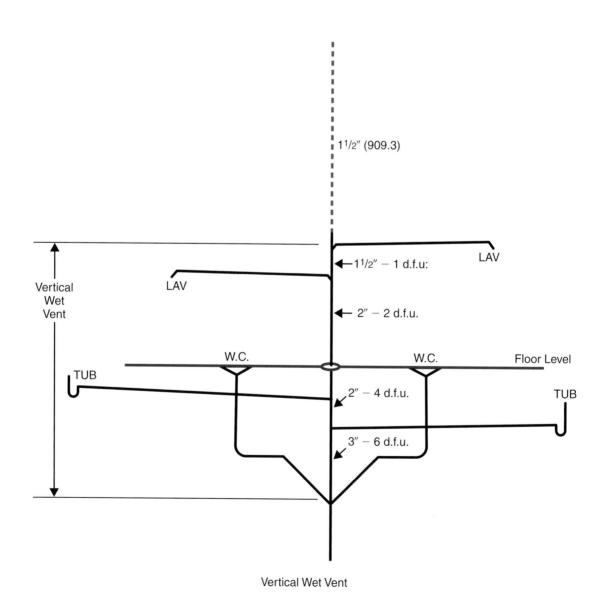

Vertical Wet Vent

909.3 Size. <u>The dry vent serving the wet-vent shall be sized based on the largest required diameter of pipe within the wet vent system served by the dry vent</u>. The wet vent shall be of a minimum size as specified in Table 909.3, based on the fixture unit discharge to the wet vent.

CHANGE SIGNIFICANCE. Section 909.1 is revised in its heading and has two text insertions to address horizontal wet vents. The horizontal reference here clearly identifies that the section regulates only horizontal wet venting.

Revisions in Section 909.1.1 further define the vertical wet application. The editorial changes make the horizontal and vertical sections more consistent. The addition of the words *wet-vented* in the third sentence is intended to make clear that only wet-vented fixture drains are required to connect independently to the vertical wet vent. Fixture drains served by dry vents are not required to connect independently to the stack, but their drainage fixture unit value loads must be considered when sizing the vertical wet vent. This independence or common connection is identified here since the practical consideration of plumbing connections on a vertical application must involve "limited height restriction." A "limited height restriction" in this context means that stacking sanitary tees to pick up water closets, tubs, or showers and then lavatories in a confined height space requires planning. The location of the stack must be considered; for example, a centered stack may require common sanitary crosses; a stack outside back-to-back restrooms may require fittings stacked independently or sanitary tees receiving common, like fixtures. In summary, installers are required to plan ahead in limited space to enable practical connections to conform to code requirements.

Revisions in Sections 909.2 acknowledge that in vertical wet vent systems the point where the dry vent connects to the wet vent is always located at the most upstream fixture drain connection to the vertical stack. The third sentence makes it clear that in horizontal wet vent systems, multiple wet-vented fixtures within the bathroom groups shall not discharge upstream of the fixture providing the vent for the system. When horizontal wet vents are installed in this manner, fixture traps upstream of the dry-vented fixture drain connection can be subjected to negative pressures. The potential for negative pressures is greatest when one of the upstream fixtures is a water closet. Also, the first sentence of 909.3 is a relocation of sizing clarification text deleted from section 909.2.

Every illustration depicting approved horizontal wet venting in the international codes commentary and in many training booklets indicates that the fixture drain providing the vent is either the most upstream fixture connection to the horizontal branch drain or the second most upstream connection. Illustrations depicting nonapproved horizontal wet venting show the fixture drains providing the vent connection further downstream than is permitted.

The Section 909.3 revision clarifies and ensures the basis for sizing the dry vent, which is the same criteria for both systems.

910.2, 910.3

Stack Installation, Stack Vent, and Another Stack Vent (Waste Stack Vents)

CHANGE TYPE. Clarification

CHANGE SUMMARY. The first section revision clarifies that vented special waste stack system areas receiving individual fixtures shall not have any offsets. The second revision clarifies that the vent for the special waste stack system shall not be reduced; however, it may connect to other vents.

2006 CODE: 910.2 Stack Installation. The waste stack shall be vertical, and both horizontal and vertical offsets shall be prohibited <u>between the lowest fixture drain connection and the highest fixture drain connection</u>. Every fixture drain shall connect separately to the waste stack. The stack shall not receive the discharge of water closets or urinals.

910.3 Stack Vent. A stack vent shall be provided for the waste stack. The size of the stack vent shall be ~~equal to~~ <u>not less than</u> the size

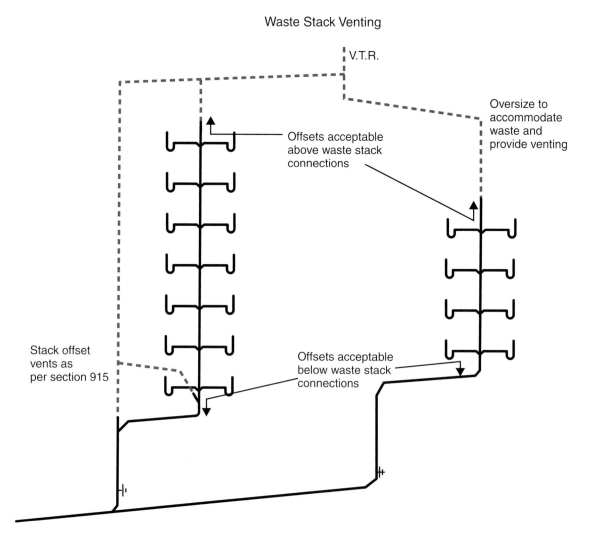

Waste Stack Venting

V.T.R.

Offsets acceptable above waste stack connections

Oversize to accommodate waste and provide venting

Stack offset vents as per section 915

Offsets acceptable below waste stack connections

Stack Installations (For Waste Stacks)

of the waste stack. Offsets shall be permitted in the stack vent, shall be located at least 6 inches (152 mm) above the flood level of the highest fixture and shall be in accordance with Section 905.2. <u>The stack vent shall be permitted to connect with other stack vents and vent stacks in accordance with Section 903.5.</u>

CHANGE SIGNIFICANCE.   The Section 910.2 revision adds clarification to the intent of this section that offsets are permitted below the lowest fixture drain connection and above the highest fixture drain connection. Conversely, as pointed out in the Change Summary, offsets are not allowed in the area receiving fixture discharges. The proponent stated that the theory is to keep open bore flow vertically because of the simultaneous flow and venting that will occur in these oversized systems. As long as vertical piping installation is maintained (no offsets, vertical or horizontal) between the lowest fixture connection and the highest fixture connection, this open bore flow will exist uninterrupted. Open bore flow ensures that amounts of waste are limited and that proper amounts of air are supplied and circulated to prevent surges of waste in the pipe.

Clear understanding of the Section 910.3 revisions provides that the vent size above the stack shall not be reduced. It further clarifies that this referenced vent may connect to other vents. The first change mandates that these connections shall not be diminished.

912.2

Installation (Section of Combination Drain and Vent)

CHANGE TYPE.   Modification

CHANGE SUMMARY.   The revision was submitted to delete three fixture references from Section 912.2 (a sink, lavatory, and drinking fountain). The proponent stated that fixtures allowed on combination drain and vent systems are already listed in Section 912.1.

2006 CODE: 912.2 Installation.   The only vertical pipe of a combination drain and vent system shall be the connection between the fixture drain ~~of a sink, lavatory or drinking fountain,~~ and the horizontal combination drain and vent pipe. The maximum vertical distance shall be 8 feet (2438 mm).

CHANGE SIGNIFICANCE.   The intent was to eliminate conflicts in the code commonly referred to as redundant language. This change will clarify that the fixtures listed in 912.1 may have a vertical component from the trap arm to the combination drain and vent system.

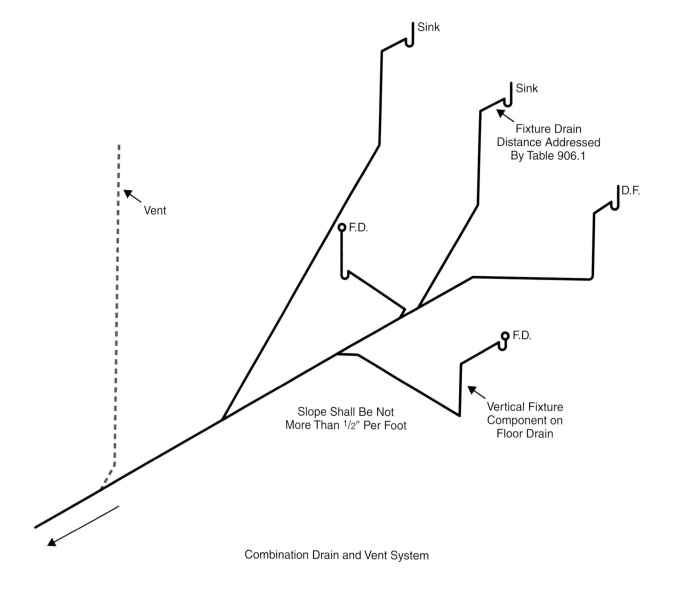

Combination Drain and Vent System

This change will also clarify that floor drains may have a vertical leg when properly installed in combination drain and vent system applications. Installations and training diagrams will now feature a vertical fixture drain component for floor drains to the horizontal portion of the combination drain and vent system.

The vertical leg is limited to 8 foot and is sized in accordance with Table 912.3. The table ensures oversizing, allowing additional air for venting from the oversized maximum slope–regulated horizontal portion of the combination drain and vent system.

1002.1
Fixture Traps

CHANGE TYPE.   Modification

CHANGE SUMMARY.   The section was revised to clarify the maximum allowable distance a trap could extend from the fixture outlet and correctly identify a clothes washer standpipe dimension as height rather than distance.

2006 CODE: 1002.1 Fixture Traps.   Each plumbing fixture shall be separately trapped by a water-seal trap, except as otherwise permitted by this code. ~~The trap shall be placed as close as possible to the fixture outlet.~~ The vertical distance from the fixture outlet to the trap weir shall not exceed 24 inches (610 mm) <u>and the horizontal distance shall not exceed 30 inches (610 mm) measured from the centerline of the fixture outlet to the centerline of the inlet of the trap.</u> The ~~distance~~ <u>height</u> of a clothes washer standpipe above a trap shall conform to Section 802.4. A fixture shall not be double trapped.

Exceptions:
1. This section shall not apply to fixtures with integral traps.
2. A combination plumbing fixture is permitted to be installed on one trap, provided that one compartment is not more than

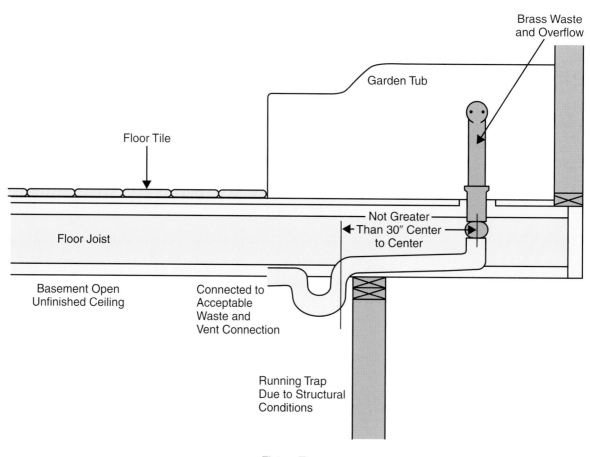

Fixture Traps

6 inches (152 mm) deeper than the other compartment and the waste outlets are not more than 30 inches (762 mm) apart.

3. A grease trap intended to serve as a fixture trap in accordance with the manufacturer's installation instructions shall be permitted to serve as the trap for a single fixture or a combination sink of not more than three compartments where the vertical distance from the fixture outlet to the inlet of the interceptor does not exceed 30 inches (762 mm) and the developed length of the waste pipe from the most upstream fixture outlet to the inlet of the interceptor does not exceed 60 inches (1524 mm).

CHANGE SIGNIFICANCE. Past code text attempted to keep the trap as close to the fixture outlet as possible. This revision will provide greater clarity and establish adequate distance limitations that are similar to the 30-inch prescribed distances in other areas of the chapter limiting trap-to-outlet distances.

Traps installed away from the outlet in the horizontal position are classified as running traps. Running traps do not have adequate self-cleansing properties as normal installations and may foul and become inoperable sooner than conventional installations. However, running trap installations are often necessary when limited installation depths are available under a fixture such as a whirlpool tub on a building overhang.

1003.1

Where Required (Interceptors and Separators)

CHANGE TYPE. Clarification

CHANGE SUMMARY. The section is revised to recognize private sewage disposal systems, which are the only alternatives to a public system.

2006 CODE: 1003.1 Where Required. Interceptors and separators shall be provided to prevent the discharge of oil, grease, sand, and other substances harmful or hazardous to the building drainage system, the public sewer, <u>the private sewage disposal system,</u> or <u>the</u> sewage treatment plant or processes.

CHANGE SIGNIFICANCE. The 2003 code did not reference private sewage disposal systems. Many commercial kitchens and food processing locations must connect to private disposal systems when public systems are not available, such as in rural areas.

The reference to private and public systems does not exclude a pump and haul system. Authorities having jurisdiction, such as health departments or sewage treatment authorities, may require the pump and haul systems.

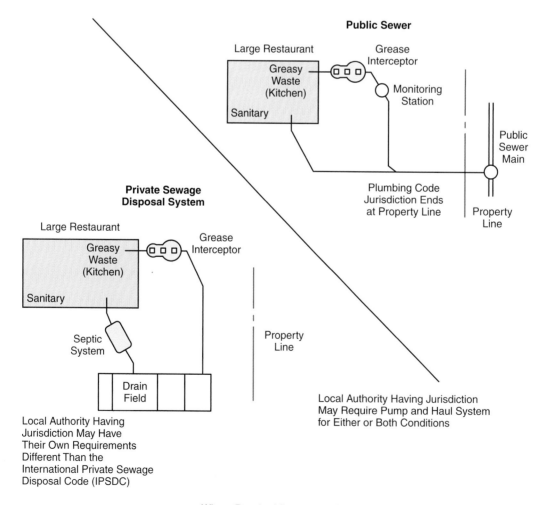

Where Required (Interceptors)

1003.3.1 and 1003.3.2

Grease Interceptors and Automatic Grease Removal Devices and Food Waste Grinders Required

CHANGE TYPE.   Modification

CHANGE SUMMARY.   Both sections are revised to insert the corrected term *grease interceptors,* to include new technology in grease removal devices, and to ensure that harmful chemical additives are not discharged to food waste grinders.

2006 CODE: 1003.3.1 Grease ~~Traps, Grease~~ Interceptors and Automatic Grease Removal Devices Required.   A ~~grease trap, or~~ grease interceptor <u>or automatic grease removal device</u> shall be required to receive the drainage from fixtures and equipment with grease-laden waste located in food preparation areas, such as in restaurants, hotel kitchens, hospitals, school kitchens, bars, factory cafeterias, and clubs. <u>Fixtures and equipment shall include pot sinks; pre-rise sinks; soup kettles or similar devices; wok stations; floor drains or sinks into which kettles are drained; automatic hood wash units and dishwashers without prerinse sinks. Grease interceptors and automatic grease removal devices shall receive waste only from fixtures and equipment that allow fats, oils, or grease to be discharged.</u>

1003.3.2 Food Waste Grinders.   Where food waste grinders connect to grease ~~traps~~ <u>interceptors</u>, a solids interceptor shall separate the discharge before connecting to the grease ~~trap~~ <u>interceptor</u>. Solids interceptors and grease interceptors shall be sized and rated for the discharge of the food waste grinder. <u>Emulsifiers, chemicals, enzymes and bacteria shall not discharge into the food waste grinder.</u>

CHANGE SIGNIFICANCE.   Section 1003.3.1 is revised to provide consistency with several other approved grease interceptor changes

1003.3.1 and 1003.3.2 continues

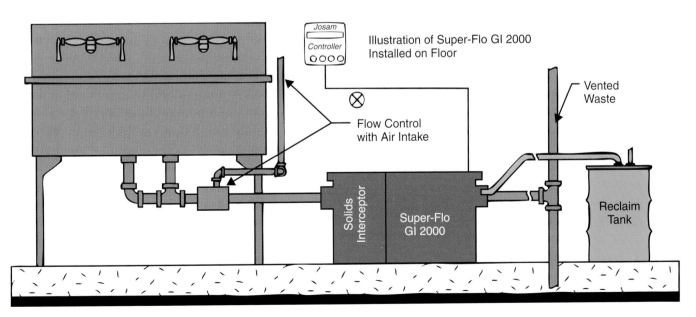

Grease Interceptors and Automatic Grease Removal Devices

1003.3.1 and 1003.3.2 continued and deletes the term *grease trap* for the corrected standard term *grease removal devices.* These devices shall only receive the discharge from fixtures requiring treatment or separation. The additional language includes the listing of fixtures for clarification examples.

Section 1003.3.2 is revised in accordance with referenced standards and should be consistent with ASME A112.14.3, which indicates that all grease traps are called interceptors. The Plumbing Drainage Institute's "Guide to Grease Interceptors" states that "the use of chemicals, often touted as environmentally friendly enzymes or emulsifiers, are to be avoided along with the use of 'bacteria' or organisms designed to digest waste." Waste from food waste grinders should be removed and not broken down in a solid collector.

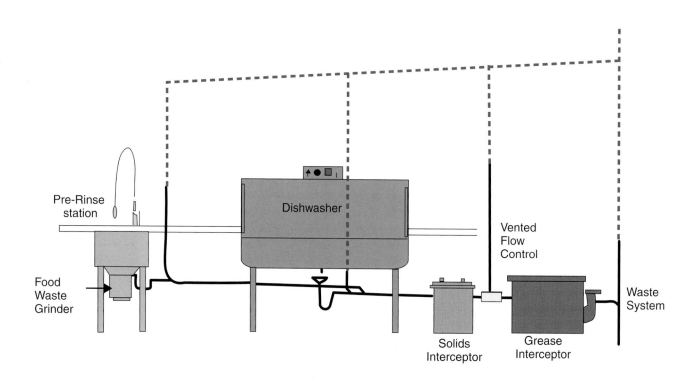

Pre-Rinse station

Dishwasher

Vented Flow Control

Food Waste Grinder

Solids Interceptor

Grease Interceptor

Waste System

1003.3, 1003.3.3, 1003.3.4, 1003.3.4.1, Table 1003.3.4.1, and 1003.3.4.2

Grease Interceptors; Grease Interceptors and Automatic Grease Removal Devices Not Required; Grease Interceptor Capacity; and Grease Interceptors and Automatic Grease Recovery Removal Devices

CHANGE TYPE. Modifications

CHANGE SUMMARY. The sections are revised to correctly identify grease interceptors rather than grease traps and discuss grease removal devices identified in their respective standards, along with two additional sizes for PDI G101 devices.

2006 CODE: 1003.3 Grease Traps and Grease Interceptors. Grease traps and grease interceptors shall comply with the requirements of Sections 1003.3.1 through 1003.3.5.

1003.3.3 Grease ~~Traps, and Grease~~ Interceptors and <u>Automatic Grease Removal Devices</u> Not Required. A grease ~~trap~~, <u>interceptor</u> or <u>automatic grease removal device</u> ~~or a grease interceptor~~ shall not be required for individual dwelling units or any private living quarters.

1003.3.4 Grease ~~Traps and Grease~~ Interceptors and <u>Automatic Grease Removal Devices.</u> Grease ~~traps and grease~~ interceptors or automatic grease <u>removal devices</u> shall conform to PDI G101, ASME A112.14.3, or ASME A112.14.4 and shall be installed in accordance with the manufacturer's instructions.

1003.3.4.1 Grease ~~Trap~~ Interceptor Capacity. Grease ~~traps~~ <u>interceptors</u> shall have the grease retention capacity indicated in Table 1003.3.4.1 for the flow-through rates indicated.

1003.3.4.2 Rate of Flow Controls. Grease ~~traps~~ <u>interceptors</u> shall be equipped with devices to control the rate of water flow so that the water flow does not exceed the rated flow. The flow-control device shall be vented and terminate not less than 6 inches (152 mm) above the flood rim level or be installed in accordance with the manufacturer's instructions.

TABLE 1003.3.4.1 Capacity of Grease ~~Traps~~ Interceptors[a]

Total Flow-Through Rating (gpm)	Grease Retention Capacity (Pounds)
4	8
6	12
7	14
9	18
10	20
12	24
14	28
15	30
18	36
20	40
25	50
35	70
50	100
<u>75</u>	<u>150</u>
<u>100</u>	<u>200</u>

a.<u>For total flow-through ratings greater than 100 (gpm), double the flow-through rating to determine the grease retention capacity (pounds).</u>

JOSAM Steel Grease Interceptor
Conforms to PDI-G101

1003.3, 1003.3.3, 1003.3.4, 1003.3.4.1, Table 1003.3.4.1, and 1003.3.4.2

continues

1003.3, 1003.3.3, 1003.3.4, 1003.3.4.1, Table 1003.3.4.1, and 1003.3.4.2 continued

CHANGE SIGNIFICANCE. Current text causes a conflict between the code and referenced standards and should be consistent with such standards, which indicate that all grease traps are now called grease interceptors and all grease recovery devices are grease removal devices.

Section 1003.3.4.1 and its table specify that traps and interceptors shall conform to the referenced standards; however, these standards pertain to devices designed to handle flows less than 100 gpm by code definition. The modification to the table is in accordance with ASME A112.14.3 for standard flow rates and grease retention capacity ratings for grease interceptors.

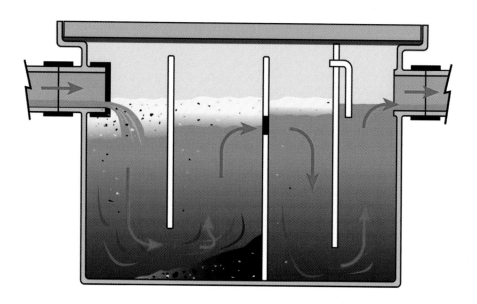

Grease Interceptor

CHANGE TYPE.   Addition

CHANGE SUMMARY.   The section is added to clarify where grease removal devices shall be used.

2006 CODE: **1003.3.5 Automatic Grease Removal Devices.** Where automatic grease removal devices are installed, such devices shall be located downstream of each fixture or multiple fixtures in accordance with the manufacturer's instructions. The automatic grease removal device shall be sized to pretreat the measured or calculated flows for all connected fixtures or equipment. Ready access shall be provided for inspection and maintenance.

CHANGE SIGNIFICANCE.   Section 1003.3.4 of the code references ASME A112.14.4, Grease Removal Device. The devices should be clarified in the code. The additional language includes the device's location and sizing requirements.

1003.3.5
Automatic Grease Removal Devices

1003.4

Oil Separators Required

CHANGE TYPE. Modification

CHANGE SUMMARY. The revisions clarify text for car wash facilities and requires system protection for hydraulic elevator pits.

2006 CODE: 1003.4 Oil Separators Required. At repair garages, car-washing facilities, ~~with engine or undercarriage cleaning capability~~ at factories where oily and flammable liquid wastes are produced <u>and in hydraulic elevator pits,</u> separators shall be installed into which all oil-bearing, grease-bearing or flammable wastes shall be discharged before emptying in the building drainage system or other point of disposal.

> **Exception:**
> <u>An oil separator is not required in hydraulic elevator pits where an approved alarm system is installed.</u>

CHANGE SIGNIFICANCE. The revision provides editorial clarification to existing text as to where an oil separator is required. All car washes have engine or undercarriage cleaning capability, so this text is redundant. Further, Section 1003.4.2.2 requires an oil separator where cars are washed.

The proposed change will require oil separators for hydraulic elevator pits. Hydraulic elevator pits have shown to leak hydraulic fluid. The exception allows a warning device permitted by ANSI A17.1 to be installed rather than oil separators.

**JOSAM
60500B-EGOLD**
Oil Interceptor with
Electronic Grease/Oil
Level Detector

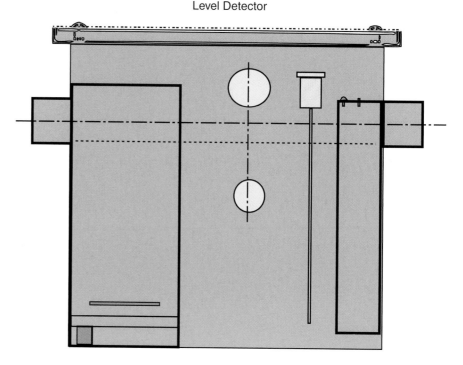

1003.6
Laundries

CHANGE TYPE.   Modification

CHANGE SUMMARY.   The section is revised to clarify that all laundries other than those in individual units shall have a means to protect the sewer from unacceptable items from the washer.

2006 CODE: 1003.6 Laundries.   ~~Commercial laundries~~ Laundry facilities not installed within an individual dwelling unit or intended for individual family use shall be equipped with an interceptor with a wire basket or similar device, removable for cleaning, that prevents passage into the drainage system of solids 0.5 inch (12.7 mm) or larger in size, string, rags, buttons, or other materials detrimental to the public sewage system.

CHANGE SIGNIFICANCE.   The revised section prevents the passage of solids to the drainage system by requiring a collection device in all laundry facilities other than an individual dwelling unit. Laundry facilities not installed within an individual dwelling unit shall now be equipped with an interceptor with a wire basket or similar device for cleaning that prevents the passage of solids. The individual who proposed the code change stated a laundry in an apartment complex is not "public" but definitely needs a method of interception based on the high volume of usage.

Lint Interceptor

1003.6 continues

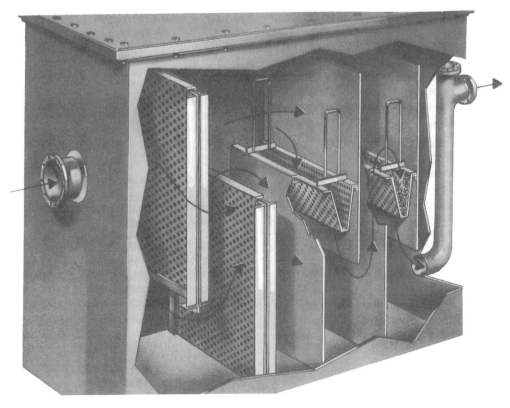

Lint Separator
(Courtesy of Rockford Sanitary Systems)

1003.6 continued

The term *commercial,* which is not defined in the code and in the past was generally applied only to self-service laundries, is removed. The phrase "not intended for individual family use" is based on the type II clothes dryer definition found in the International Mechanical Code. Large institutional laundry machines commonly are drained by gravity out the back into a large drain, where the collection device could easily be installed.

This code change will be an installation challenge. For example, many machines—other than single-family-dwelling machines, which discharge to a 2″ standpipe—now appear to need a collection device. Finding or creating room for these devices on the same floor level will be difficult.

CHANGE TYPE.   Modification

CHANGE SUMMARY.   Table 1102.4 is revised to include two additional CSA standards for acrylonitrile butadiene styrene storm sewer plastic pipe and an additional CSA standard for polyvinyl chloride storm sewer plastic pipe. Table 1102.5 is revised to include three CSA standards for polyethylene plastic pipe.

Table 1102.4 and Table 1102.5

Building Storm Sewer Pipe and Sub-Soil Drain Pipe

2006 CODE:

TABLE 1102.4   **Building Storm Sewer Pipe**

Material	Standard
Acrylonitrile butadiene styrene (ABS) plastic pipe	CSA B181.1, CSA B182.1
Polyvinyl chloride (PVC) plastic pipe (Type DWV, SDR26, SDR35, SDR41, PS50, PS100)	~~CAN~~/CSA B182.4, CSA B181.2

TABLE 1102.5   **Subsoil Drain Pipe**

Material	Standard
Polyethylene (PE) plastic pipe	CSA B182.1; CSA B182.6; CSA B182.8
Polyvinyl chloride (PVC) Plastic pipe (type sewer pipe, PS25, PS50 or PS100)	~~CAN~~/CSA B182.4

*Reader's Note: *Portions of tables not covered in this book as significant changes will remain as printed from the 2003 code edition.*

CHANGE SIGNIFICANCE.   The building storm sewer pipe acceptance table, 1102.4, now includes three additional plastic pipe CSA standards. The CSA standards for acrylonitrile butadiene styrene and polyvinyl chloride storm plastic pipe are recognized for sanitary applications in Chapter 7.

The subsoil drain pipe acceptance table, 1102.5, now includes three additional plastic pipe CSA standards. The three newly recognized polyethylene plastic pipe standards will join the presently listed ASTM F405 standard in the table.

The code revisions enable manufacturers with products meeting the applicable CSA or ASTM standards to have their products used and accepted by the authorities having jurisdiction.

Table 1102.7

Pipe Fittings

CHANGE TYPE. Additions

CHANGE SUMMARY. The table was revised to correct which pipe fitting standards apply to storm sewer applications. Several were added and others removed.

TABLE 1102.7 Pipe Fittings

Material	Standard
Acrylonitrile butadiene styrene (ABS) plastic	~~ASTM D2468~~; ASTM D2661; ASTM D3311; CSA B181.1
~~Chlorinated Polyvinyl chloride (CPVC) plastic~~	~~ASTM F437; ASTM F438; ASTM F439~~
~~Polyethylene (PE) plastic~~	~~ASTM D2609~~
Polyvinyl chloride (PVC) plastic	~~ASTM D2464; ASTM D2466; ASTM D2467; CSA CAN/CSA B137.2~~; ASTM D2665; ASTM D 3311; ASTM F1866
Coextruded composite ABS DWV Schedule 40 IPS pipe (solid or cellular core)	ASTM D2661; ASTM D3311; ASTM F628
Coextruded composite PVC DWV Schedule 40 IPS-DR, PS140, PS200 (solid cellular)	ASTM D2665; ASTM D3311; ASTM F891
Coextruded composite ABS sewer and drain DR-PS in PS35, PS50, PS100, PS140, PS200	ASTM D2751
Coextruded composite PVC sewer and drain DR-PS in PS35, PS50, PS100, PS140, PS200	ASTM D3034

*Reader's Note: *Portions of tables not covered in this book as significant changes will remain as printed from the 2003 code edition.*

CHANGE SIGNIFICANCE. The revisions to the building storm sewer pipe fittings acceptance table, 1102.7, removed standards addressing water supply and distribution materials and added fittings acceptable for drain and waste fitting from other areas of the code. A water piping material example would be chlorinated polyvinyl chloride plastic, meeting standards ASTM F437, ASTM F438, and ASTM F439, which are now deleted. A drain and waste fitting example would be polyvinyl chloride plastic, meeting standards ASTM D 3311, which is now added.

Chapter 13
Referenced Standards

CHANGE TYPE. Modifications

CHANGE SUMMARY. Numerous presently listed standards have been revised in recognition of newer published editions of the standard, and many new standards have been added. These changes occurred over two separate code update processes from the 2003 to 2006 code editions.

Reader's Note: *The following two standard references are examples of the many updates that have been approved in the code change process.*

ANSI, American National Standards Institute

25 West 43rd Street

Fourth Floor

New York, NY 10036

Standard reference number: Z21.22-~~1999~~ (R2003)

Title: Relief Valves for Hot Water Supply Systems with Addenda Z21.22a-~~2000~~ (R2003) and Z21.22b-2001 (R2003)

NFPA, National Fire Protection Association

1 Batterymarch Park

Quincy, MA 02269–9101

Standard reference number: 99C-02 ~~99~~

Title: Gas and Vacuum Systems

CHANGE SIGNIFICANCE. The plumbing code documents for both cycles of the 2006 updates referenced over 180 standards changes. The two examples above indicate changes considered for relief valves conforming to ANSI standards and NFPA medical gas piping standards. Numerous other standards have been updated and added. The two above are used as examples for discussion purposes.

The standards listed in Chapter 13 are referenced in other portions of the code for enforcement purposes. The referenced standards are updated through consensus processes. The content of the individual standards is of extreme importance and at times involves extensive changes from one standard edition to the next.

The *ICC Code Development Process for the International Codes* (Procedures) Section 4.5 requires the updating of referenced standards to be accomplished administratively and to be processed as code proposals administratively by the appropriate code development committee. A letter was sent to each developer of standards that were referenced in the International Code Council codes, asking them to provide a list of their standards in order to update to the current edition.

Jurisdictions adopting codes are generally required to participate in a legal process that allows code review and public comment on the specific language and (at times) amendments found in the code being considered. Following adoption, the specific language referenced in

Chapter 13 continues

Chapter 13 continued

the document will prevail. Flexibility for the jurisdiction process, such as code officials' decisions and variances, are established in Chapter 1 of the code. However, abiding by the specific edition of the published standard is often mandated. Conflicts occur when code officials are compelled to adhere to published standards editions, whereas manufactures and designers are required to conform to the most recent standards edition.

In summary, the referenced standards found in the code are an important part of the code requirements. The plumbing code benefits from the uniformity of consensus standards rather than relying on recommendations from individual manufacturers. When differences occur between the referenced standards and the code, the code supersedes the standards.

Appendix C
Gray Water Recycling Systems

CHANGE TYPE. Modification

CHANGE SUMMARY. The revised Appendix C text expands information to include in building use specifications for outside irrigation applications and inside flushing systems. Existing text was deleted.

APPENDIX C GRAY WATER RECYCLING SYSTEMS

Note: Section 301.3 of this code requires all plumbing fixtures that receive water or waste to discharge to the sanitary drainage system of the structure. In order to allow for the utilization of a gray water recycling system, Section 301.3 should be revised to read as follows:

301.3 Connections to Drainage System. All plumbing fixtures, drains, appurtenances, and appliances used to receive or discharge liquid wastes or sewage shall be directly connected to the <u>sanitary</u> drainage system of the building or premises, in accordance with the requirements of this code. This section shall not be construed to prevent indirect waste systems <u>required by</u> ~~provided for in~~ Chapter 8.

Exception:
Bathtubs, showers, lavatories, clothes washers and laundry sinks shall not be required to discharge to the sanitary drainage system where such fixtures discharge to an approved gray water recycling system.

***Reader's Note:** All of the following existing sections have been deleted with their complete text, as printed in the new code. The section headings have been listed here for clarification.*

Appendix C continues

Appendix C continued   **C101**

~~**GRAY WATER RECYCLING SYSTEMS**~~

~~**C101.1 General** . . .~~

~~**C101.2 Definition** . . .~~

~~**GRAY WATER** . . .~~

~~**C101.3 Installation** . . .~~

~~**C101.4 Reservoir** . . .~~

~~**C101.5 Filtration** . . .~~

~~**C101.6 Disinfections** . . .~~

~~**C101.7 Makeup water** . . .~~

~~**C101.8 Overflow** . . .~~

~~**C101.9 Drain** . . .~~

~~**C101.10 Vent required** . . .~~

~~**C101.11 Coloring** . . .~~

~~**C101.12 Identification** . . .~~

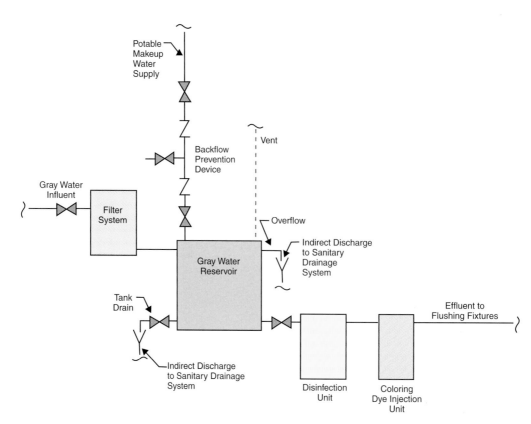

*Reader's Note: *All of the following sections have been added with their complete text, as printed in the new code. The section headings have been listed here for clarification.*

SECTION C101 GENERAL

C101.1 Scope . . .

C101.2 Definition. The following term shall have the meaning shown herein.

GRAY WATER. Waste discharged from lavatories, bathtubs, showers, clothes washers, and laundry trays.

C101.3 Permits . . .

C101.4 Installation . . .

C101.5 Materials . . .

C101.6 Tests . . .

C101.7 Inspections . . .

C101.8 Potable water connections . . .

C101.9 Waste water connections . . .

C101.10.1 Required valve . . .

C101.11 Collection reservoir . . .

C101.12 Overflow . . .

C101.13 Drain . . .

C101.14 Vent required . . .

SECTION C102 SYSTEMS FOR FLUSHING WATER CLOSETS AND URINALS

C102.1 Collection reservoir . . .

102.2 Disinfections . . .

102.3 Makeup water . . .

102.4 Coloring . . .

C102.5 Materials . . .

C102.6 Identification . . .

Appendix C continues

TABLE C 103.10 Distribution Pipe . . .

FIGURE 1 Subsurface Landscape Irrigation (see page 123)

CHANGE SIGNIFICANCE. Appendix C has been revised to delete existing text and add new wording, which will provide greater clarification and now includes a definition of gray water, systems for flushing water closets and urinals, and subsurface landscape irrigation systems.

Conservation needs and concerns are growing internationally. In order to meet those concerns, gray water recycling systems will become more desirable and possibly mandated. The past provisions for gray water recycling systems included requirements for both residential and commercial applications. Further, previous provisions did not specifically address irrigation reuse, specification of materials, and sizing requirements for irrigation purposes.

The new code language was based on a combination of practices utilized in the International Private Sewage Disposal Code, ASPE Data Book, Gray Water Guidelines published by the Water Conservation Association of Southern Arizona, and California health laws related to recycled water.

PART 2

International Mechanical Code

Chapters 1 Through 15

The *International Mechanical Code*® (IMC) contains provisions for the regulation of mechanical equipment design and installation. The code consists of 15 chapters and two appendices. Appendix A deals with combustion air openings and chimney connector pass-throughs and Appendix B lists recommended permit fees.

The provisions of Chapter 1 address the application, enforcement, and administration of subsequent requirements of the code. Chapter 2 provides definitions for terms used throughout the IMC. Chapter 3 includes the general requirements for listed equipment, appliance location, protection for personnel servicing mechanical equipment, access requirements for appliances in various locations, and condensate disposal. Chapter 4 addresses building ventilation. Chapter 5 includes exhaust systems such as commercial kitchen exhaust systems. Chapter 6 covers all of the duct construction for HVAC systems. Chapter 7 addresses combustion air for solid and liquid fueled appliances. Chapter 8 has requirements for vents and chimneys. Chapter 9 includes requirements for specific appliances, fireplaces, and solid-fuel equipment. Chapter 10 deals with boilers, water heaters, and pressure vessels. Chapter 11 addresses refrigeration systems. Chapter 12 has requirements for hydronic piping. Regulations governing fuel oil piping and storage are located in Chapter 13. Chapter 14 deals with solar systems. Chapter 15 contains standards, referenced in the 2006 IMC, listed by the promulgating agency of the standard. ■

301.4

Listed and Labeled

CHANGE TYPE. Clarification

CHANGE SUMMARY. The change to the 2006 edition of the International Mechanical Codes (IMC) clarifies that a listed product shall be used as stated and intended in the listing.

2006 CODE: 301.4 Listed and Labeled. ~~All appliances~~ Appliances regulated by this code shall be listed and labeled <u>for the application in which they are installed and used,</u> unless otherwise approved in accordance with Section 105.

> **Exception:** <u>Listing and labeling of equipment and applications used for refrigeration shall be in accordance with Section 1101.2.</u>

CHANGE SIGNIFICANCE. This section has been modified to bring to the attention of the code official, engineer, architect, or installing contractor that just being listed doesn't necessarily mean that a product has been listed and tested for all applications. It is the responsibility of the engineer or architect, the code official, and the installer of a listed product to read the manufacturer's installation instructions and be familiar with the particulars of the listing on that product.

Some examples of products that have been listed but are sometimes not installed per their intended use are fire dampers, exhaust fans, duct wraps, and firestopping materials.

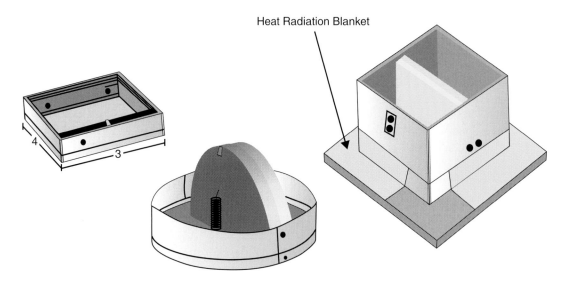

Heat Radiation Blanket

Ceiling dampers are listed for use in duct penetrations in the ceiling portion of a rated floor/ceiling or roof/ceiling assembly.

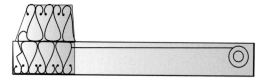

Horizontal fire damper listed for floor assemblies but not listed for use in a rated ceiling assembly.

CHANGE TYPE.   Addition

CHANGE SUMMARY.   The new Section 304.3.1 in the 2006 IMC will allow a fuel-fired appliance to be installed at the grade level of a parking garage if it is located in a room with a vestibule and two doors.

2006 CODE: 304.3. Elevation of Ignition Source.   Equipment and appliances having an ignition source and located in hazardous locations and public garages, private garages, repair garages, automotive motor-fuel-dispensing facilities and parking garages shall be elevated such that the source of ignition is not less than 18 inches (457 mm) above the floor surface on which the equipment or appliance rests. Such equipment and appliances shall not be installed in Use Group H occupancies or control areas where open use, handling, or dispensing of combustible, flammable or explosive materials occurs. For the purpose of this section, rooms, or spaces that are not part of the living space of a dwelling unit and that communicate directly with a private garage through openings shall be considered to be part of the private garage.

304.3.1 Parking Garages.   Connection of a parking garage with any room in which there is a fuel-fired appliance shall be by means of a vestibule providing a two-doorway separation, except that a single door is permitted where the sources of ignition in the appliance are elevated in accordance with Section 304.3.

> **Exception:** This section shall not apply to appliance installations complying with Section 304.5.

CHANGE SIGNIFICANCE.   This change was made to Section 304.3 in the IMC to coordinate it with Section 406.2.8 in the International Building Code (IBC). As part of the special use and occupancy re-

304.3.1 continues

304.3.1
Parking Garages

Plan View

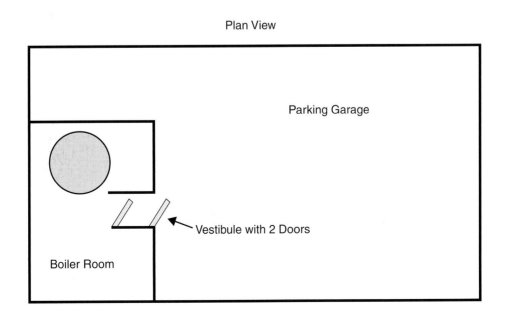

Parking Garage

Vestibule with 2 Doors

Boiler Room

304.3.1 continued   quirements for parking garages, all possible ignition sources must be controlled and isolated. Specifically, all heating equipment must be located in rooms that are separated from the main areas where the vehicles are parked. Doorways connecting the heating equipment rooms and the main parking area must have a vestibule or airlock arrangement such that one must pass through two doors prior to entering the other room. Again, this is done to minimize the possibility of any spilled flammable liquids and the resulting vapors from coming into contact with the ignition sources of the heating equipment.

An allowance is made if the heating equipment is located at least 18 inches (457 mm) above the floor of the separated room. In such a case, the vestibule/airlock arrangement with double door system is not required and can involve just a single door. If this exception is used, care must be taken by the building and fire officials that these special stipulations and conditions are part of the certificate of occupancy.

Elevation

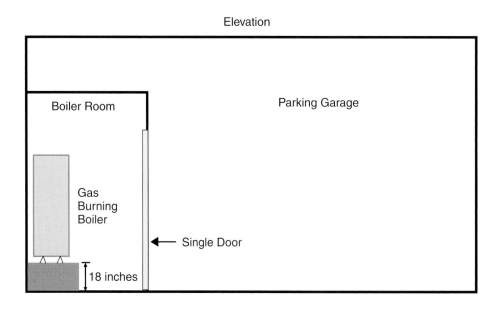

CHANGE TYPE. Addition

304.10
Guards

CHANGE SUMMARY. Section 304.10 in the 2006 IMC has been changed by adding roof hatch openings to the requirements for guards.

2006 CODE: 304.10 Guards.   Guards shall be provided where appliances, equipment, fans, or other components that require service and <u>roof hatch openings</u> are located within 10 feet (3048 mm) of a roof edge or open side of a walking surface and such edge or open side is located more than 30 inches (762 mm) above the floor, roof or grade below. The guard shall extend not less than 30 inches (762 mm) beyond each end of such appliance<u>s</u>, equipment, fans, component<u>s</u>, and <u>roof hatch openings</u> and the top of the guard shall be located not less than 42 inches (1067 mm) above the elevated surface adjacent to the guard. The guard shall be constructed so as to prevent the passage of a 21-inch-diameter (533 mm) sphere and shall comply with the loading requirements for guards specified in the *International Building Code.*

CHANGE SIGNIFICANCE.   The IMC has always required guards on the service sides of mechanical equipment in elevated locations. Section 304.10 will now require that guards be installed adjacent to roof hatch openings that are closer than 10 feet to an open side that is more than 30 inches above a floor, roof, or grade level. The roof hatches are used by service personnel, inspectors, emergency responders, and others. The roof hatches are frequently used during inclement weather, emergency situations, and darkness. For all of these

304.10 continues

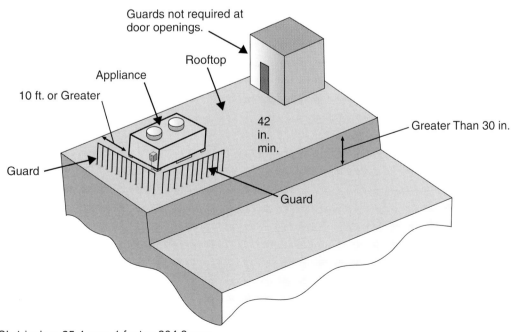

Guards not required at door openings.

Rooftop

Appliance

10 ft. or Greater

42 in. min.

Greater Than 30 in.

Guard

Guard

For SI: 1 inch = 25.4 mm, 1 foot = 304.8 mm.

304.10 continued   reasons, the guards are necessary to provide the same minimum measure of safety for roof hatch openings that is provided at the service side of the mechanical equipment. It should be noted that this addition to Section 304.10 was intended to include roof hatch openings only. Other types of roof access openings such as doors do not require guards.

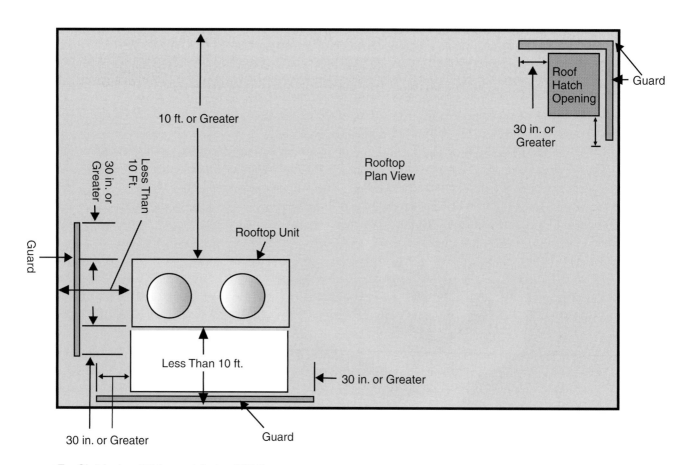

For SI: 1 inch = 25.4 mm, 1 foot = 304.8 mm.

CHANGE TYPE. Addition

CHANGE SUMMARY. The code will now allow a passageway up to 50 feet in an attic that is at least 6 feet high and 22 inches wide.

2006 CODE: 306.3 Appliances in Attics. Attics containing appliances requiring access shall be provided with an opening and unobstructed passageway large enough to allow removal of the largest appliance. The passageway shall not be less than 30 inches (762 mm) high and 22 inches (559 mm) wide and not more than 20 feet (6096 mm) in length measured along the center line of the passageway from the opening to the appliance. The passageway shall have continuous solid flooring not less than 24 inches (610 mm) wide. A level service space not less than 30 inches (762 mm) deep and 30 inches (762 mm) wide shall be present at the front or service side of the appliance. The clear access opening dimensions shall be a minimum of 20 inches by 30 inches (508 mm by 762 mm), where such dimensions are large enough to allow removal of the largest appliance.

306.3 continues

306.3
Appliances in Attics

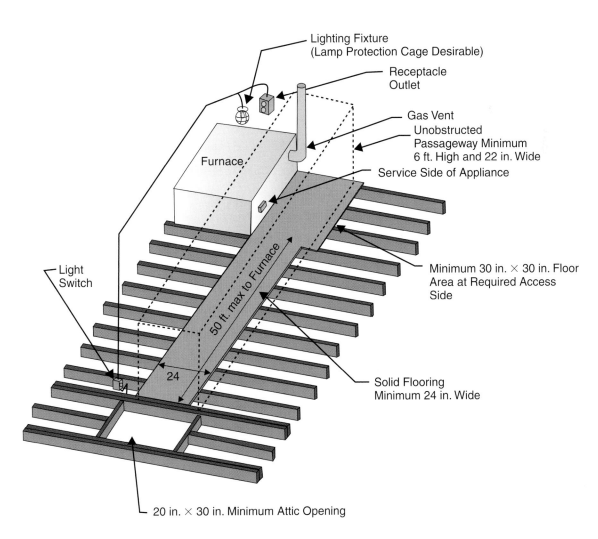

Attic Furnace Installation

306.3 continued

Exceptions:

1. The passageway and level service space are not required where the appliance is capable of being serviced and removed through the required opening.

2. Where the passageway is unobstructed and not less than 6 feet (1829 mm) high and 22 inches wide for its entire length, the passageway shall be not greater than 50 feet (15,250 mm) in length.

CHANGE SIGNIFICANCE. The new Exception 2 to Section 306.3 in the 2006 IMC will allow for an extended travel distance to appliances in attics. This distance may now extend up to 50 feet, measured along the center line of the passageway, provided the attic is not less than 6 feet high and 22 inches wide for its entire length. Prior codes required a 20-foot maximum length for the passageway, regardless of the height or width of the attic space. This change was submitted to correlate this section in the IMC with Section M306 in the International Fuel Gas Code (IFGC). The new exception recognizes that the 6-foot-high passage will allow for better access and that most individuals that will have to use this space to service the equipment in the attic space will be able to do so safely and easier when there is a greater height along the passageway.

CHANGE TYPE. Addition

CHANGE SUMMARY. The code will now allow a passageway of unlimited length in an underfloor space that is at least 6 feet high and 22 inches wide.

2006 CODE: 306.4 Appliances Under Floors. Underfloor spaces containing appliances requiring access shall be provided with an access opening and unobstructed passageway large enough to remove the largest appliance. The passageway shall not be less than 30 inches (762 mm) high and 22 inches (559 mm) wide, nor more than 20 feet (6096 mm) in length measured along the centerline of the passageway from the opening to the appliance. A level service space not less than 30 inches (762 mm) deep and 30 inches (762 mm) wide shall be present at the front or service side of the appliance. If the depth of the passageway or the service space exceeds 12 inches (305 mm) below the adjoining grade, the walls of the passageway shall be lined with concrete or masonry. Such concrete or masonry shall extend a minimum of 4 inches (102 mm) above the adjoining grade and shall have sufficient lateral-bearing capacity to resist collapse. The clear access opening dimensions shall be a minimum of 22 inches by 30 inches (559 mm by 762 mm), where such dimensions are large enough to allow removal of the largest appliance.

Exceptions:
1. The passageway is not required where the level service space is present when the access is open and the appliance is capa-

306.4 continues

306.4
Appliances Under Floors

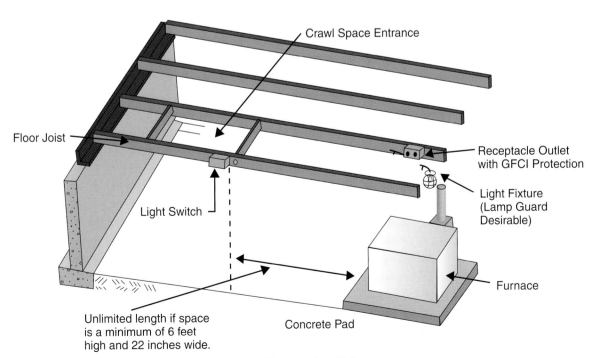

Crawl Space Furnace Installation

Crawl Space Entrance

Floor Joist

Receptacle Outlet with GFCI Protection

Light Fixture (Lamp Guard Desirable)

Light Switch

Furnace

Unlimited length if space is a minimum of 6 feet high and 22 inches wide.

Concrete Pad

306.4 continued

ble of being serviced and removed through the required opening.

2. <u>Where the passageway is unobstructed and not less than 6 feet high (1929 mm) and 22 inches wide for its entire length, the passageway shall not be limited in length.</u>

CHANGE SIGNIFICANCE.   The new Exception 2 to Section 306.4 in the 2006 IMC will allow for an extended travel distance to appliances in underfloor areas. The length will now be unlimited, provided the underfloor space is not less than 6 feet high and 22 inches wide for its entire extent. Prior codes required a 20-foot maximum length of the passageway, regardless of the height or width of the underfloor space.

This change was submitted to correlate this section in the IMC with Section M306.4 in the IFGC. The codes recognize that when access to mechanical equipment in an underfloor area is a minimum of 6 feet high and 22 inches wide, there is greater area for servicing the equipment and a better and quicker means of egress in the case of an emergency. It is similar to accessing equipment in any other common room or space.

CHANGE TYPE. Clarification

CHANGE SUMMARY. The 2003 Section 306.6 in the IMC has been assigned to 306.5.1, and it has been made clear that the guard rails on level service platforms on sloped roofs are required only when such equipment needs service.

2006 CODE: ~~306.6~~ 306.5.1 Sloped Roofs. Where appliances, <u>equipment, fans, or other components that require service</u> are installed on a roof having a slope of 3 units vertical in 12 units horizontal (25% slope) or greater and having an edge more than 30 inches (762 mm) above grade at such edge, a level platform shall be provided on each side of the appliance to which access is required ~~by the manufacturer's installation instructions~~ for service, repair, or maintenance. The platform shall be not less than 30 inches (762 mm) in any dimension, and shall be provided with guards. ~~in accordance with Section 304.16.~~ <u>The guards shall extend not less than 42 inches above the platform, shall be constructed so as to prevent the passage of a 21-inch (533 mm) diameter sphere and shall comply with the loading requirements for guards specified in the *International Building Code.*</u>

306.5.1 continues

306.5.1
Sloped Roofs

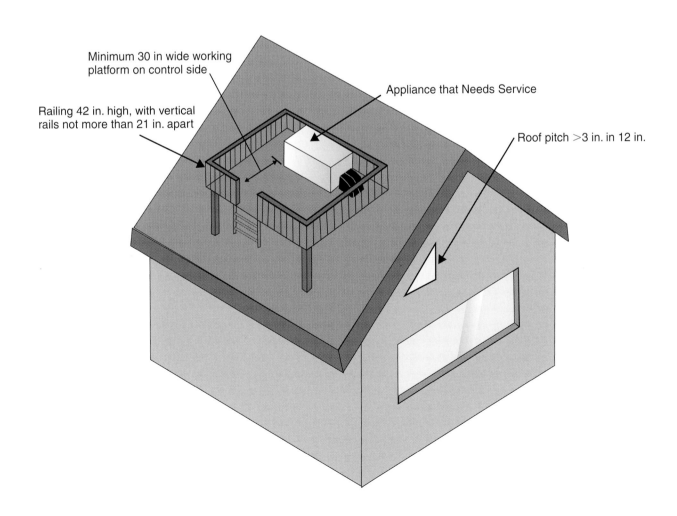

Minimum 30 in wide working platform on control side

Railing 42 in. high, with vertical rails not more than 21 in. apart

Appliance that Needs Service

Roof pitch >3 in. in 12 in.

306.5.1 continued   **CHANGE SIGNIFICANCE.**   The requirements for guards around service panels of mechanical equipment in elevated locations have always been in the code. It was never clear whether these requirements applied to level service platforms on sloped roofs and whether they applied to all equipment, regardless of service needs. The hazards to service personnel are the same, regardless of whether the equipment is installed on a flat roof or a service platform of a sloped roof. The possibility of someone falling or service equipment being knocked off onto the ground and injuring someone are the same in both applications. The revisions in this section clarify that guards are required in all elevated locations where service of equipment is needed.

CHANGE TYPE.   Addition

CHANGE SUMMARY.   The code will now allow for water level detectors in locations that were not previously permitted.

2006 CODE: 307.2.3 Auxiliary and Secondary Drain Systems. In addition to the requirements of Section 307.2.1, a secondary drain or auxiliary drain pan shall be required for each cooling or evaporator coil <u>or fuel-fired appliance that produces condensate,</u> where damage to any building components will occur as a result of overflow from the equipment drain pan or stoppage in the condensate drain piping. One of the following methods shall be used:

1. An auxiliary drain pan with a separate drain shall be provided under the coils on which condensation will occur. The auxiliary pan drain shall discharge to a conspicuous point of disposal to alert occupants in the event of a stoppage of the pri-

307.2.3 continues

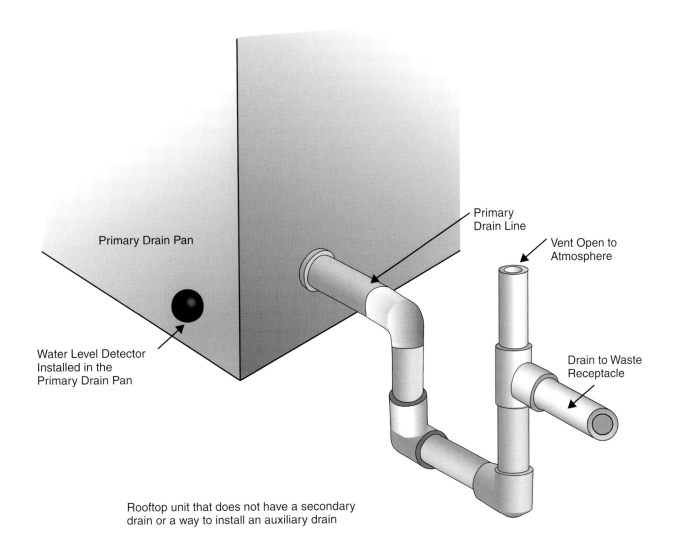

Primary Drain Pan

Primary Drain Line

Vent Open to Atmosphere

Water Level Detector Installed in the Primary Drain Pan

Drain to Waste Receptacle

Rooftop unit that does not have a secondary drain or a way to install an auxiliary drain

307.2.3 continued

mary drain. The pan shall have a minimum depth of 1.5 inches (38 mm), shall not be less than 3 inches (76 mm) larger than the unit or the coil dimensions in width and length and shall be constructed of corrosion-resistant material. Metallic pans shall have a minimum thickness of not less than 0.0276-inch (0.7 mm) galvanized sheet metal. Nonmetallic pans shall have a minimum thickness of not less than 0.0625 inch (1.6 mm).

2. A separate overflow drain line shall be connected to the drain pan provided with the equipment. Such overflow drain shall discharge to a conspicuous point of disposal to alert occupants in the event of a stoppage of the primary drain. The overflow drain line shall connect to the drain pan at a higher level than the primary drain connection.

3. An auxiliary drain pan without a separate drain line shall be provided under the coils on which condensate will occur. Such pan shall be equipped with a water-level detection device conforming to UL 508 that will shut off the equipment served prior to overflow of the pan. The auxiliary drain pan shall be constructed in accordance with Item 1 of this section.

4. A water level detection device conforming to UL 508 shall be provided that will shut off the equipment served in the event that the primary drain is blocked. The device shall be installed in the primary drain line, the overflow drain line, or in the equipment-supplied drain pan, located at a point higher than the primary drain line connection and below the overflow rim of such pan.

Exception: Fuel-fired appliances that automatically shut down operation in the event of a stoppage in the condensate drainage system.

CHANGE SIGNIFICANCE. The new Item 4 to Section 307.2.3 in the 2006 IMC will allow for water level detectors to be used at the primary drain line, at the overflow drain line, or in the equipment-supplied drain pan. The water level detector may be used in lieu of a separate overflow drain line.

The use of a water level detector instead of a secondary condensate drain line may result in an easier and safer installation. The code requires that a secondary condensate line shall discharge at a conspicuous point of disposal. This is an undefined statement and can lead to some points of disposal that may not be desirable. If the secondary condensate line terminated through the ceiling in the middle of a room it would be in a conspicuous location, but if it started to drain because of a restricted primary condensate line, then a computer or other valuable equipment or furniture beneath it might be damaged. The area under the secondary condensate line could be a path of egress or another walking area where someone could slip and fall if the condensate water was on the floor. In some installations, the space above the ceiling is very small and there is not enough height to get the proper slope for a secondary condensate line to terminate in a safe place, such as through the ceiling over a lavatory, where if it started to drain and was not detected right away it would not do any damage.

With the new code requirements that will allow for the water detectors in lieu of the secondary condensate lines, these possible problems will be eliminated.

307.2.3.1

Water Level/ Monitoring Devices

CHANGE TYPE. Addition

CHANGE SUMMARY. The code will now require water level detectors on down-flow units.

2006 CODE: <u>**307.2.3.1 Water Level Monitoring Devices.** On down-flow units and all other coils that do not have a secondary drain and do not have a means to install an auxiliary drain pan, a water level monitoring device shall be installed inside the primary drain pan. This device shall shut off the equipment served in the event that the primary drain becomes restricted. Externally installed devices and devices installed in the drain line shall not be permitted.</u>

CHANGE SIGNIFICANCE. The new Section 307.2.3.1 in the 2006 IMC will now require water level detectors on down-flow units that do not have a secondary drain or where an auxiliary drain pan cannot be installed. The detector shall shut off the equipment and alert the occupants if the primary drain line is restricted. The water level detector will have to be installed in the unit. External devices will not be al-

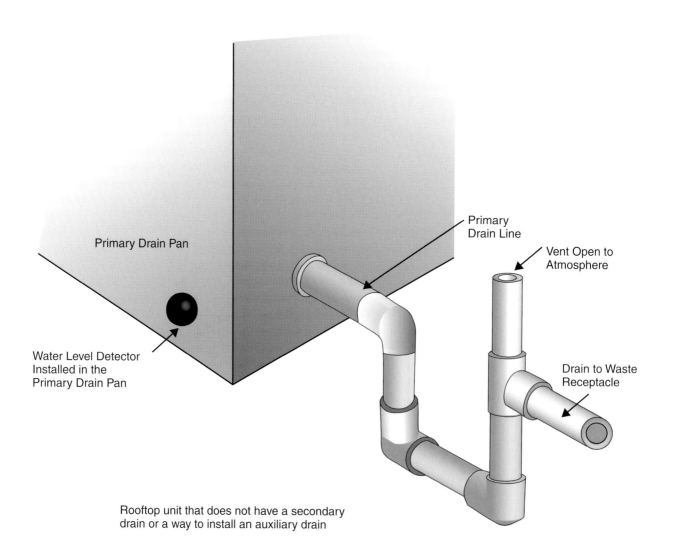

Primary Drain Pan

Primary Drain Line

Vent Open to Atmosphere

Drain to Waste Receptacle

Water Level Detector Installed in the Primary Drain Pan

Rooftop unit that does not have a secondary drain or a way to install an auxiliary drain

lowed. It should be noted that this new section in the 2006 IMC was intended to include packaged rooftop cooling units.

If the primary condensate drain line on a down-flow <u>or rooftop</u> cooling unit is restricted, the condensate will accumulate in the pan provided by the equipment manufacturer. If this happens, the condensate will create mold and mildew in the unit. The condensate will also leak out through the seams of the condensate pan and may overflow into the duct system. This can cause damage to the duct insulation and to the building and its contents. The water level detector will shut off the equipment and alert the occupants that there is a problem with the primary condensate line.

401.4.1

Intake Openings

CHANGE TYPE.   Clarification

CHANGE SUMMARY.   This addition to Section 401.4.1 in the 2006 IMC will allow an exhaust outlet from a bathroom or kitchen in a residential dwelling to terminate closer than 10 feet horizontally or 2 feet below a mechanical or gravity outdoor air-intake opening.

2006 CODE: ~~**401.5.1**~~ **401.4.1 Intake Openings.**   Mechanical and gravity outdoor air intake openings~~,~~ shall be located a minimum of 10 feet (3048 mm) <u>horizontally</u> from any hazardous or noxious contaminant <u>source,</u> such as vents, chimneys, plumbing vents, streets, alleys, parking lots, and loading docks, except as otherwise specified in this code. Where a source of contaminant is located within 10 feet (3048 mm) <u>horizontally</u> of an intake opening, such opening shall be located a minimum of 2 feet (610 mm) below the contaminant source.

<u>The exhaust from a bathroom or kitchen in a residential dwelling shall not be considered to be a hazardous or noxious contaminant.</u>

CHANGE SIGNIFICANCE.   This revision clarifies that bathrooms and kitchens in residential dwellings do not emit vapors that would be considered to be hazardous or noxious to the point of causing physical harm to individuals or the surrounding structure. In many jurisdictions, the "10 foot/2 foot" rule is not enforced in apartment or motel applications because the limited wall areas do not provide enough space to achieve the 10 foot spacing and because the emissions have never been considered noxious by most code officials. It should be

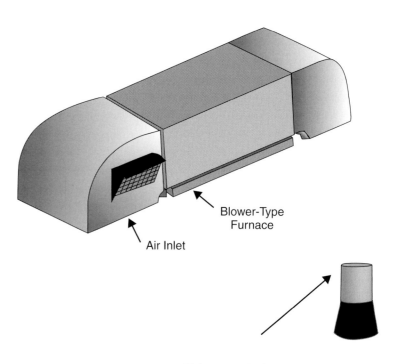

Exhaust outlet from a kitchen or bathroom in a residential dwelling. Does not have to terminate 10 feet away or 2 feet above outside air inlet.

noted that residential bathroom and kitchen exhausts are considered environmental air exhaust and need to comply with Section 401.4. It should also be noted that this section mentions bathroom and kitchen exhaust in a residential dwelling only. For other environmental exhaust terminations such as bathrooms or toilet rooms in commercial buildings and clothes dryers, the requirements in Section 501.2.1 in the 2006 IMC would apply.

403.2.1

Recirculation of Air

Table 403.3

Required Outdoor Ventilation Air (see page 232)

CHANGE TYPE. Addition

CHANGE SUMMARY. The additions to Section 403.2.1 Items 2 and 4 will allow ventilation air from previously prohibited areas to be recirculated back into the building.

2006 CODE: **403.2.1 Recirculation of Air.** The air required by Section 403.3 shall not be recirculated. Air in excess of that required by Section 403.3 shall not be prohibited from being recirculated as a component of supply air to building spaces, except that:

1. Ventilation air shall not be recirculated from one dwelling to another or to dissimilar occupancies.

2. Supply air to a swimming pool and associated deck areas shall not be recirculated unless such air is dehumidified to maintain the relative humidity of the area at 60% or less. Air from this area shall not be recirculated to other spaces <u>where 10% or</u>

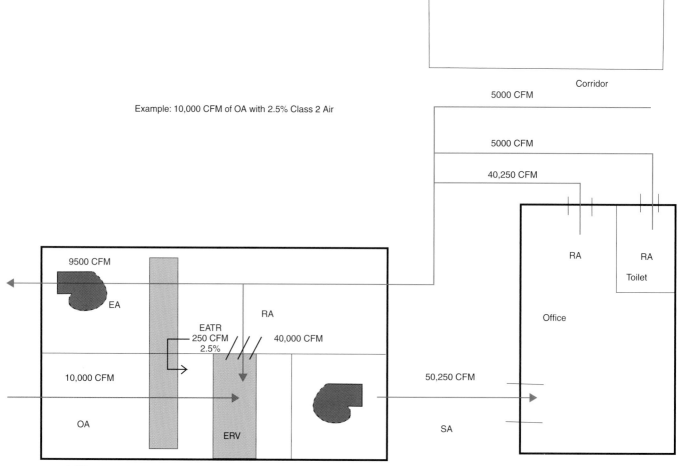

Example: 10,000 CFM of OA with 2.5% Class 2 Air

EATR = Exhaust Air Through Recoverer

more of the resulting supply airstream consists of air recirculated from these spaces.

3. Where mechanical exhaust is required by <u>Note b in</u> Table 403.3, recirculation of air from such spaces shall be prohibited. All air supplied to such spaces shall be exhausted, including any air in excess of that required by Table 403.3.

4. <u>Where mechanical exhaust is required by Note h in Table 403.3, mechanical exhaust is required and recirculation is prohibited where 10% or more of the resulting supply airstream consists of air recirculated from these spaces.</u>

CHANGE SIGNIFICANCE. The additions to Section 403.2.1 originated with some changes that took place with the ASHRAE 62 Standard. What this new wording does is allow exhaust air from areas like swimming pools, bathrooms, toilet rooms, and locker rooms to be recirculated to other parts of a building. The code based on the ASHRAE 62 Standard does limit the recirculated air from these areas to not more than 10%. The recirculated air is ducted through an energy recovery ventilation (ERV) system. The ERVs have tremendous potential for energy savings because they are designed to reduce heat energy losses in the heating season and reduce heat energy gains in the cooling season. The explanations include examples of what this code change will allow.

■ Classifies the "quality" of air you will be exhausting from the space and using for energy recovery. Controlling recirculation is the goal.

■ *Class 1: Air with low contaminant concentration*—office classrooms, conference room, etc.

403.2.1 continues

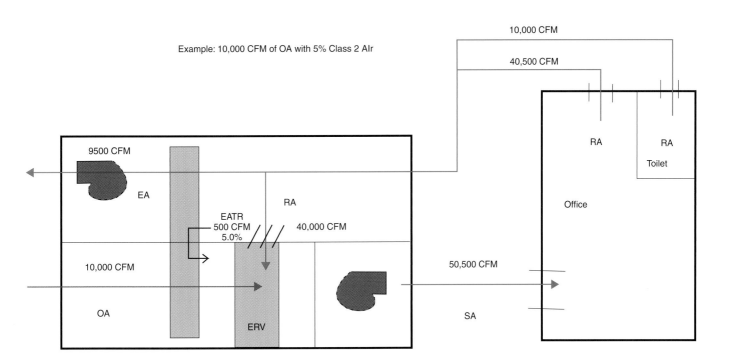

403.2.1 continued

- *Class 2: "Air with moderate contaminant concentration, mildly offensive odors or sensory-irritation intensity"*—toilet exhaust, locker room, dry cleaning, commercial laundry, etc.

- *Class 3: "Air with significant contaminant concentration or significant offensive odor or sensory-irritation intensity"*—daycare sickroom, general chemical/biological labs, machinery rooms, etc.

- *Class 4: "Air with highly objectionable fumes or gases or potentially containing dangerous particles, bioaerosols, or gases at a concentration high enough to be considered harmful"*—laboratory hoods, diazo machine exhaust, paint spray booth, etc.

- Example: Class 2 Air

  - A 10,000-cfm air handler providing 100% outside air to an office, using a draw-through enthalpy wheel with cross-leakage of 5% EATR (exhaust air transfer ratio); exhaust is taken from the toilet rooms (class 2 air)

    - If all exhaust is from the restrooms: Supply air will be 5% class 2 air; it may be redesignated as class 1 and may be supplied to any space.

    - If 5000 cfm comes from the restrooms and 5000 cfm from the corridors, supply air will be 2.5% class 2 air and may be supplied to any space.

See page 232 for Table 403.3: Required Outdoor Ventilation Air.

CHANGE TYPE.   Addition

CHANGE SUMMARY.   All of the requirements for the termination of exhaust ducts will be in one section. There are now requirements for the termination of exhaust ducts that were not in the previous codes.

2006 CODE: 501.2 ~~Outdoor~~ Exhaust Discharge.   The air removed by every mechanical exhaust system shall be discharged outdoors at a point where it will not cause a nuisance and <u>not less than the distances specified in Section 501.2.1. The air shall be discharged to a location</u> from which it cannot again be readily drawn in by a ventilating system. Air shall not be exhausted into an attic or crawl space.

Exceptions:
1. Whole-house ventilation-type attic fans ~~that~~ <u>shall be permitted to</u> discharge into the attic space of dwelling units having private attics.
2. Commercial cooking recirculating systems.

501.2.1 Location of Exhaust Outlets.   <u>The termination point of exhaust outlets and ducts discharging to the outdoors shall be located with the following minimum distances:</u>

1. <u>For ducts conveying explosive or flammable vapors, fumes or dusts: 30 feet (9144 mm) from property lines; 10 feet (3048 mm) from operable openings into buildings; 6 feet (1829 mm)</u>

501.2.1 continues

501.2.1

Location of Exhaust Outlets

Exhaust Ducts Conveying Explosive or
Flammable Vapors Fumes or Dusts

501.2.1 continued

from exterior walls and roofs; 30 feet (9144 mm) from combustible walls and operable openings into buildings which are in the direction of the exhaust discharge; 10 feet (3048 mm) above adjoining grade.

2. For other product-conveying outlets: 10 feet (3048 mm) from the property lines; 3 feet (914 mm) from exterior walls and roofs; 10 feet (3048 mm) from operable openings into buildings; 10 feet (3048 mm) above adjoining grade.

3. For environmental air duct exhaust: 3 feet (914 mm) from property lines; 3 feet (914 mm) from operable openings into buildings for all occupancies other than Group U, and 10 feet (3048 mm) from mechanical air intakes.

4. For specific systems: for clothes dryer exhaust, see Section 504.4; for kitchen hoods, see Section 506.3; for dust, stock and refuse conveying systems, see Section 511.2; and for subslab soil exhaust systems, see Section 512.4.

[F]502.7.3.6 Termination Point. ~~The termination point for exhaust ducts discharging to the atmosphere shall be located with the following minimum distances:~~

1. ~~For ducts conveying explosive or flammable vapors, fumes or dusts: 30 feet (9144 mm) from the property line; 10 feet (3048 mm) from openings into the building; 6 feet (1829 mm) from exterior walls and roofs; 30 feet (9144 mm) from combustible walls or openings into the building which are in the direction of the exhaust discharge; 10 feet (3048 mm) above adjoining grade.~~

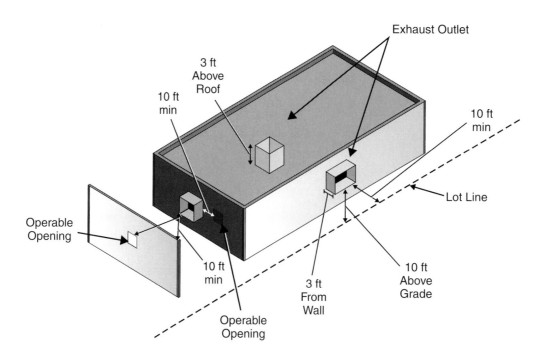

Exhaust Outlet

3 ft Above Roof

10 ft min

10 ft min

Lot Line

Operable Opening

10 ft min

3 ft From Wall

10 ft Above Grade

Operable Opening

Product Conveying Ducts That Are Not
Exhausting Explosive or Flammable Vapors
Fumes or Dusts

~~2. For other product-conveying outlets: 10 feet (3048 mm) from the property line; 3 feet (914 mm) from exterior walls and roofs; 10 feet (3048 mm) from openings into the building; 10 feet (3048 mm) above adjoining grade.~~

~~3. For environmental air duct exhaust: 3 feet (914 mm) from the property line; 3 feet (914 mm) from openings into the building.~~

511.2 Exhaust Outlets. Outlets for exhaust that exceed 600°F (315°C) shall be designed as a chimney in accordance with Table 511.2.

~~The termination point for exhaust ducts discharging to the atmosphere shall not be less than the following:~~

~~1. Ducts conveying explosive or flammable vapors, fumes or dusts: 30 feet (9144 mm) from property line; 10 feet (3048 mm) from openings into the building; 6 feet (1829 mm) from exterior walls or roofs; 30 feet (9144 mm) from combustible walls or openings into the building which are in the direction of the exhaust discharge; and 10 feet (3048 mm) above adjoining grade.~~

~~2. Other product-conveying outlets: 10 feet (3048 mm) from property line; 3 feet (914 mm) from exterior wall or roof;~~

501.2.1 continues

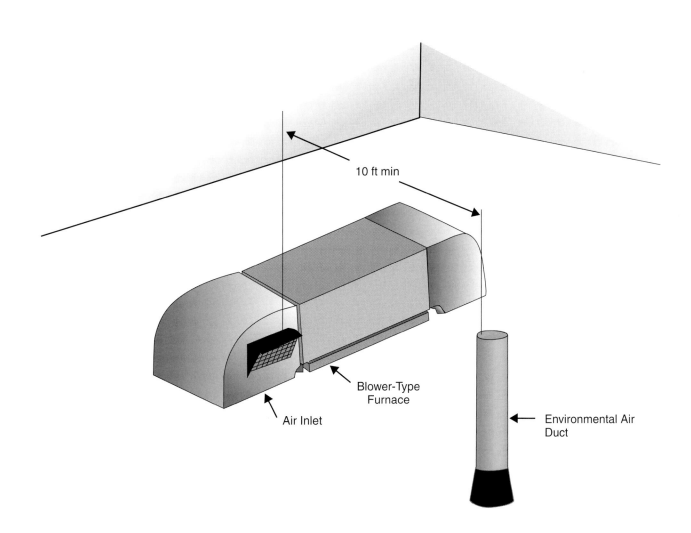

10 ft min

Blower-Type
Furnace

Air Inlet

Environmental Air
Duct

501.2.1 continued

~~10 feet (3048 mm) from openings into the building; and 10 feet (3048 mm) above adjoining grade.~~

~~**3.** Environmental air duct exhaust: 3 feet (914 mm) from property line; and 3 feet (914 mm) from openings into the building.~~

CHANGE SIGNIFICANCE.   All of the requirements for the termination of exhaust ducts can now be found in one section in the IMC. The requirements for the termination of exhaust ducts that convey flammable or explosive vapors, fumes, or dust have been relocated from Section 502.7, which deals only with the application of flammable finishes. This new section in the 2006 IMC will include the termination of exhaust ducts for all systems that convey explosive or flammable discharge and will now have requirements for the termination of product-conveying ducts. These requirements were not in the previous codes. Product-conveying ducts are intended to include exhaust systems for parking garages; fume hoods that do not convey explosive or flammable vapors, fumes, or dusts; and ducts that convey solid particles and other products that are not explosive or flammable. This was not intended to include commercial cooking systems. The required distance of 10 feet from the exhaust outlets of product-conveying systems to property lines, to operable openings into buildings, and above the adjoining grade level was intended to provide a safe distance so that fumes or odors from the exhaust system would not enter back into a building or cause a danger or hazard to the public. Item 3 in Section 501.2.1 in the 2006 IMC has requirements for environmental air duct terminations. Environmental air is now defined in Chapter 2 in the 2006 IMC as follows:

"Air which is conveyed to or from occupied areas through ducts which are not part of the heating or air-conditioning systems, such as ventilation for human usage, domestic kitchen range exhaust, bathroom exhaust, and domestic clothes dryer exhaust."

The requirements in Item 3 in Section 501.2.1 have also been relocated from Section 502.4. The new section added a requirement that an environment air duct shall terminate 10 feet from mechanical air intakes. This was added to the 2006 IMC so that the fumes from environmental air ducts would not enter back into the building through mechanical air inlets such as outside air inlets in heating and cooling systems.

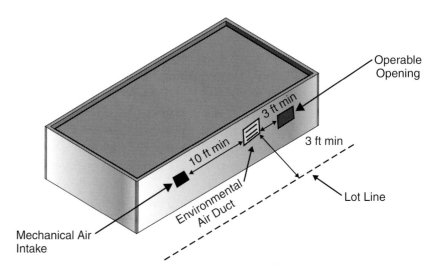

CHANGE TYPE. Modification

CHANGE SUMMARY. The code will now require a precise method of providing make-up air.

2006 CODE: 501.3 Pressure Equalization. Mechanical exhaust systems shall be sized to remove the quantity of air required by this chapter to be exhausted. The system shall operate when air is required to be exhausted. Where mechanical exhaust is required in a room or space in other than occupancies in R-3, such space shall be maintained with a neutral or negative pressure. If a greater quantity of air is supplied by a mechanical ventilating supply system than is removed by a mechanical exhaust ~~system~~ for a room, adequate means shall be provided for the natural <u>or mechanical exhaust</u> of the excess air supplied. If only a mechanical exhaust system is installed for a room or if a greater quantity of air is removed by a mechanical exhaust system than is supplied by a mechanical ventilating supply system for a room, adequate ~~means shall be provided for the natural supply of the deficiency in the air supplied~~ <u>make-up air consisting of supply air, transfer air or outdoor air shall be provided to satisfy the deficiency. The calculated building infiltration rate shall not be utilized to satisfy the requirements of this section.</u>

501.3 continues

501.3
Pressure Equalization

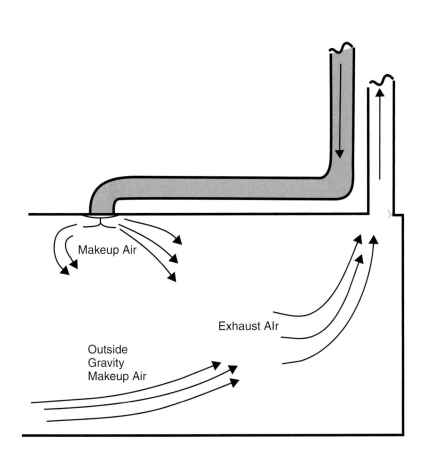

501.3 continued

CHANGE SIGNIFICANCE. The previous editions of the IMC required make-up air to be provided by a natural supply to areas that required exhaust-only systems. This was not a clear requirement for make-up air and led users of the code to believe that make-up air could be supplied only by infiltration or by gravity through permanent openings such as windows or doors.

The code is now more precise about make-up air requirements and allows for make-up air to consist of supply air, transfer air, or outdoor air.

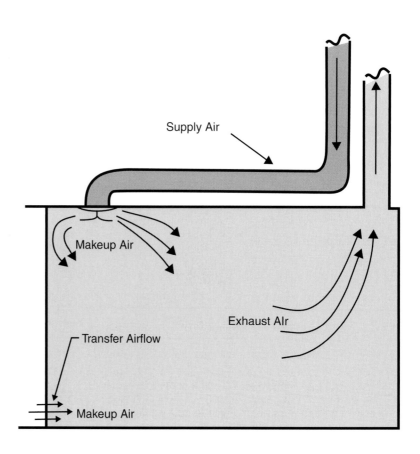

CHANGE TYPE. Modification

CHANGE SUMMARY. The change will allow a lesser velocity of 500 feet per minute within a grease duct system that serves a Type I hood.

2006 CODE: **506.3.4 Air Velocity.** Grease duct systems serving a Type I hood shall be designed and installed to provide an air velocity within the duct system of not less than ~~1500~~ 500 feet per minute (~~7.6~~ 2.5 m/s)

Exception: (No change to current text)

CHANGE SIGNIFICANCE. The modification to Section 506.3.4 will allow the velocity in a grease duct system to be not less than 500 feet

506.3.4 continues

506.3.4
Air Velocity

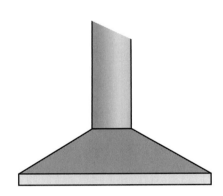

Hood is required to move a minimum of 3200 CFMs. If the minimum velocity in the grease duct is 500 feet per minute, the duct would be sized as follows: 3200 ÷ 500 × 144 = 921.6 square inches of duct area. The square root of 921.6 is 30.35. The size of the grease duct would have to be roughly 30 × 30 inches.

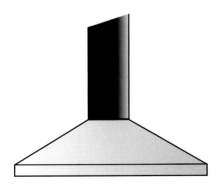

Hood is required to move a minimum of 3200 CFMs. If the minimum velocity in the grease duct is 1500 feet per minute, the duct would be sized as follows: 3200 ÷ 1500 × 144 = 307.2 square inches of duct area. The size of the grease duct would have to be roughly 17.5 × 17.5 inches.

506.3.4 continued

per minute. This has been reduced from 1500 feet per minute in the previous codes. This change was based on a recent American Society of Heating, Refrigerating, and Air-Conditioning Engineers (ASHRAE) research project (RP-1033) and will keep the IMC current with the National Fire Protection Association (NFPA) 96 Standard. It should be noted that allowing the lesser velocity in the grease duct system will necessitate an increase in the size of the duct. As an example, a 4 × 8 hood that has a gas under-fired broiler, a gas open-burner range, and a gas deep fryer under it will require a minimum airflow of 3200 cfm. If the minimum air velocity in the grease duct serving this hood is 1500 feet per minute, the duct would have to be 307.2 square inches in area; this would compute to a 17.5 × 17.5 square duct. If the minimum velocity in the duct, serving the same hood, is 500 feet per minute, then the area of the duct would increase to 921.6 square inches, which would be a 30.35 × 30.35 square duct.

CHANGE TYPE. Addition

CHANGE SUMMARY. This new section to the 2006 IMC will require automatic activation of a commercial cooking hood whenever the cooking appliances under the hood are activated.

2006 CODE: <u>507.2.1.1 Operation.</u> <u>Type I hood systems shall be designed and installed to automatically activate the exhaust fan whenever cooking operations occur. The activation of the exhaust fan shall occur through an interlock with the cooking appliances, by means of heat sensors, or by means of other approved methods.</u>

CHANGE SIGNIFICANCE. This new section to the 2006 IMC was added to ensure that a commercial cooking hook will be in operation whenever the cooking appliances under the hood are being used. When the cooking appliances under the hood are activated they will automatically activate the exhaust fan that is serving the commercial cooking hood. This will be accomplished by interlocking the cooking appliances and the exhaust fan with heat sensors or by other approved methods.

507.2.1.1
Operation

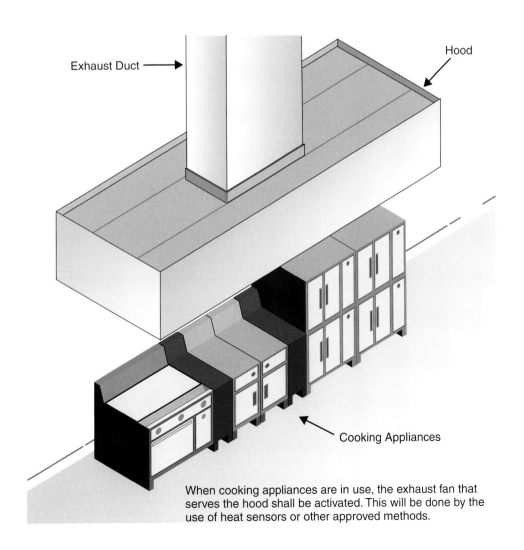

Exhaust Duct

Hood

Cooking Appliances

When cooking appliances are in use, the exhaust fan that serves the hood shall be activated. This will be done by the use of heat sensors or other approved methods.

507.2.2

Type II Hoods

CHANGE TYPE. Modification

CHANGE SUMMARY. The two exceptions that have been added to Section 507.2.2 in the 2006 IMC will clarify that certain electric cooking appliances will not require a Type II hood.

2006 CODE: 507.2.2. Type II Hoods. Type II hoods shall be installed where cooking or dishwashing appliances produce heat, ~~or~~ steam, <u>or products of combustion</u> and do not produce grease or smoke, such as steamers, kettles, pasta cookers, and dishwashing machines.

Exceptions:

1. Under-counter-type commercial dishwashing machines.
2. A Type II hood is not required for dishwashers and potwashers that are provided with heat and water vapor exhaust systems that are supplied by the appliance manufacturer and are installed in accordance with the manufacturer's instructions.
3. <u>A single light-duty electric convection, bread, retherm or microwave oven. The additional heat and moisture loads generated by such appliances shall be accounted for in the design of the HVAC system.</u>
4. <u>A Type II hood is not required for the following electrically heated appliances: toasters, steam tables, popcorn poppers, hot dog cookers, coffee makers, rice cookers, egg cookers, holding/warming ovens. The additional heat and moisture</u>

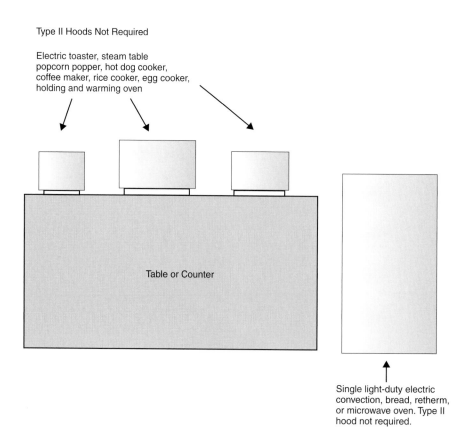

Type II Hoods Not Required

Electric toaster, steam table popcorn popper, hot dog cooker, coffee maker, rice cooker, egg cooker, holding and warming oven

Table or Counter

Single light-duty electric convection, bread, retherm, or microwave oven. Type II hood not required.

loads generated by such appliances shall be accounted for in
the design of the HVAC system.

CHANGE SIGNIFICANCE. The two new exceptions to Section
507.2.2 in the 2006 edition have exempted the requirement for a Type
II hood over small electric cooking appliances that do not produce
large amounts of heat. Both Exceptions 3 and 4 do require that any additional heat and moisture loads generated by the appliances be accounted for in the heating and cooling design. The type of appliances
that have been exempt from the requirement of a Type II hood are electric light-duty convection, bread, rethermalization (retherm), or microwave ovens, electric toasters, steam tables, popcorn poppers, hot
dog cookers, coffee makers, rice cookers, egg cookers, and holding/
warming ovens. This list was expanded from the previous codes to
agree with the ASHRAE 154 Standard Ventilation for Commercial
Cooking Operations.

510.6.1

Fire Dampers

CHANGE TYPE. Addition

CHANGE SUMMARY. The new Section 510.6.1 in the 2006 IMC will clarify that fire dampers are not to be installed where hazardous exhaust ducts penetrate fire-resistance-rated assemblies.

2006 CODE: <u>**510.6.1 Fire Dampers.**</u> <u>Fire dampers are prohibited in hazardous exhaust ducts.</u>

CHANGE SIGNIFICANCE. This section was in the 1996 and 1998 IMC but was somehow eliminated when the fire and smoke damper requirements were moved to the IBC. These IBC requirements are referenced in the IMC, but with this change the requirements for fire dampers in hazardous exhaust ducts were lost. Where fire dampers are installed in a duct system, they create an obstruction in the sys-

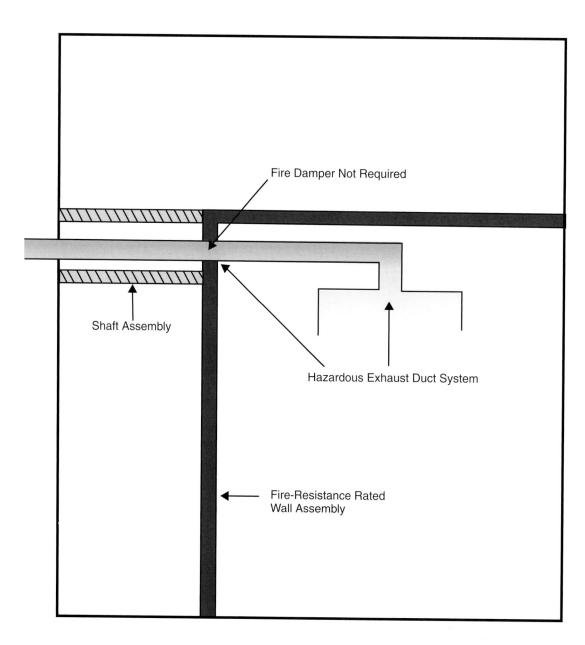

Fire Damper Not Required

Shaft Assembly

Hazardous Exhaust Duct System

Fire-Resistance Rated Wall Assembly

tem. In the case of a hazardous exhaust system, the product being exhausted may adhere or stick to the fire damper and not be exhausted to the exterior of the building. If this happens, it may create a greater hazard within the exhaust duct system. For this reason, the IMC has required that when a hazardous exhaust duct penetrates a rated floor/ceiling assembly or a fire-resistance-rated wall assembly, the duct shall be enclosed in a shaft or fire-rated construction from the point of penetration to the outlet terminal. The new section 510.6.1 clarifies that when this is done, fire dampers are not required at these duct penetrations.

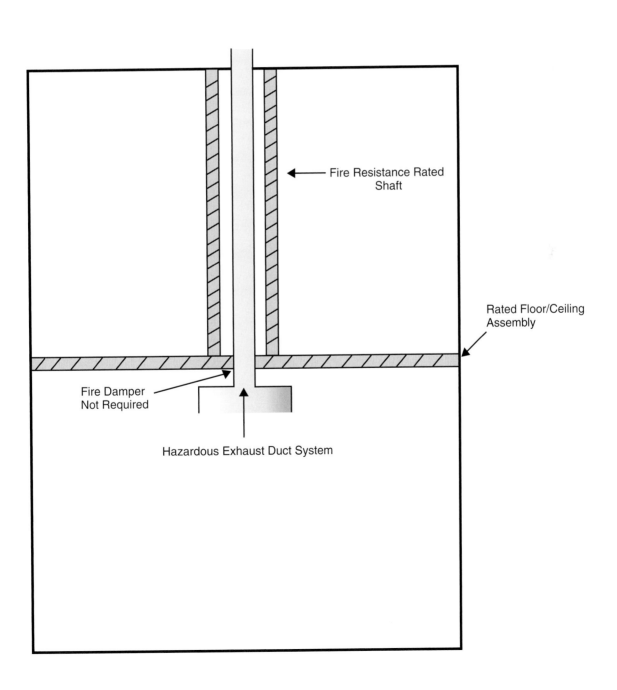

602.2.1.6

Semiconductor Fabrication Areas

CHANGE TYPE.   Addition

CHANGE SUMMARY.   This new Section 602.2.1.6 will allow underfloor and other furred spaces in a group H-5 fabrication area to be used as a plenum and not meet the 25–50 smoke and frame spread requirements in Section 602.2.1.

2006 CODE: 602.2.1 Materials Exposed within Plenums.   Except as required by Sections 602.2.1.1 through 602.2.1.5, materials exposed within plenums shall be noncombustible or shall have a flame spread index of not more than 25 and a smoke-developed index of not more than 50 when tested in accordance with ASTM E 84.

Exceptions:
1. Rigid and flexible ducts and connectors shall conform to Section 603.

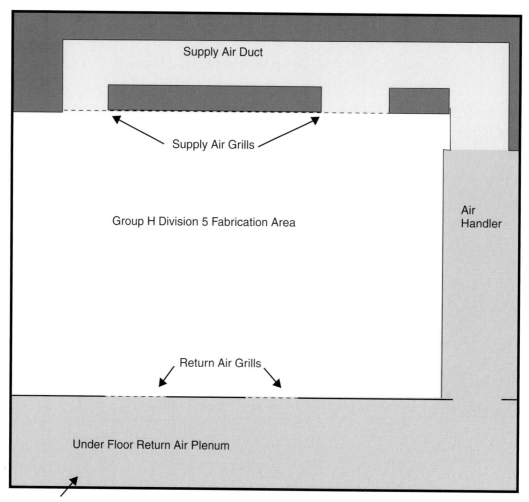

Supply Air Duct

Supply Air Grills

Group H Division 5 Fabrication Area

Air Handler

Return Air Grills

Under Floor Return Air Plenum

This space does not have to comply with Section 602.2.1 in the 2006 IMC. Material in the plenum does not have to be noncombustible or meet the 25–50 flame and smoke ratings.

2. Duct coverings, linings, tape and connectors shall conform to Sections 603 and 604.

3. This section shall not apply to materials exposed within plenums in one- and two-family dwellings.

4. This section shall not apply to smoke detectors.

5. Combustible materials enclosed in <u>noncombustible raceways or enclosures,</u> approved gypsum board assemblies or enclosed in materials listed and labeled for such application.

602.2.1.6 Semiconductor Fabrication Areas. <u>Group H, Division 5 fabrication areas and the areas above and below the fabrication area that share a common air recirculation path with the fabrication area shall not be subject to the provisions of Section 602.2.1.</u>

CHANGE SIGNIFICANCE. The spaces below and above a fabrication area (clean room) are part of a single air circulation path that operates at a high velocity. Smoke generated in these furred spaces will quickly disperse through the normally occupied areas. The plenums serving those areas are not allowed to be connected to ducts or plenums serving other areas or occupancies in the building. There is no benefit in restricting the materials in the furred spaces used as an air plenum since they are already allowed in the fabrication areas.

603.9

Joint Seams and Connections

CHANGE TYPE.   Modification

CHANGE SUMMARY.   The new addition to Section 603.9 in the 2006 IMC will specify how joints in the duct system are to be sealed and that tapes, mastics, and mechanical fasteners for use with flexible ducts shall be marked to show compliance with the appropriate UL standard.

2006 CODE: 603.9 Joint Seams and Connections.   All longitudinal and transverse joints, seams, and connections in metallic and nonmetallic ducts shall be constructed as specified in SMACNA HVAC *Duct Construction Standards—Metal and Flexible* and ~~SMACNA *Fibrous Glass Duct Construction Standards* or~~ NAIMA *Fibrous Glass Duct Construction Standards.* ~~All longitudinal and transverse joints, seams and connections shall be sealed in accordance with the *International Energy Conservation Code.*~~ <u>All joints, longitudinal and transverse seams, and connections in ductwork shall be securely fastened and sealed with welds, gaskets, mastics (adhesives), mastic-plus-embedded-fabric systems or tapes. Tapes and mastics used to seal ductwork listed and labeled in accordance with UL 181A shall be marked "181A-P" for pressure-sensitive tape, "181 A-M" for mastic or "181 A-H" for heat-sensitive tape. Tapes and mastics used to seal flexible air ducts and flexible air connectors shall comply with UL 181B and shall be marked "181B-FX" for pressure-sensitive tape or "181B-M" for mastic. Duct connections to flanges of air distribution system equipment shall be sealed and mechanically fastened. Mechanical fasteners for use with flexible nonmetallic air ducts shall</u>

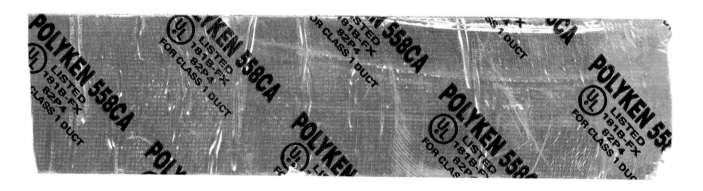

comply with UL 181B and shall be marked 181B-C. Unlisted duct tape is not permitted as a sealant on any metal ducts.

CHANGE SIGNIFICANCE. Section 603.9 now clarifies how longitudinal and transverse seams and connections in duct system are to be sealed. The code specifies that duct systems shall be sealed with welds, gaskets, mastics, mastic-plus-embedded-fabric systems, or tapes. Tapes and mastics used with UL-listed flex ducts are required to be marked on the product so that the installer and inspector know that this particular tape or mastic complies with the appropriate UL standards. The code also states that unlisted duct tape is not permitted as a sealant on any metal duct. The tape and mastics used with flex ducts have been tested with the product and are part of the UL 181-A and 181-B standards. The UL 181-B Standard has also included mechanical fasteners that are used with nonmetallic flex ducts. When the mechanical fasteners, tapes, and mastics are exposed to high temperatures for long periods of time, they can break and deteriorate. These duct systems are commonly installed in attics, where the temperatures can reach well above 120°F in the warmer months. If the tapes, mastics, and mechanical fasteners fail when they are exposed to the high temperatures, it can cause the ducts to leak or separate, causing an inefficient heating or cooling system. The requirements in this section of the code will ensure that the methods of sealing a duct system have been tested and are part of a listed system. Mechanical fasteners for flexible nonmetallic air ducts shall comply with UL 181B and shall be marked 181B-C.

Mechanical fasteners for flexible nonmetalic air ducts shall comply with UL 181 B and shall be marked 181B–C.

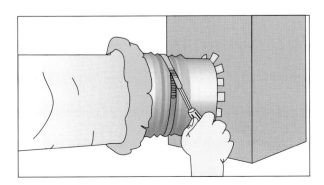

603.17

Registers, Grilles, and Diffusers

CHANGE TYPE. Addition

CHANGE SUMMARY. The addition to Section 603.17 in the 2006 IMC will require volume dampers to be provided with access.

2006 CODE: **603.17 Registers, Grilles, and Diffusers.** Duct registers, grilles, and diffusers shall be installed in accordance with the manufacturer's installation instructions. ~~Balancing~~ Volume dampers or other means of supply air adjustment shall be provided in the branch ducts or at each individual duct register, grille, or diffuser. Each volume damper or other means of supply air adjustment used in balancing shall be provided with access.

CHANGE SIGNIFICANCE. The previous codes required volume dampers in either the branch ducts or at each individual duct register. There have been problems with some installations where the adjusting mechanism for the volume damper is not accessible and the air flow to that particular area or space cannot be adjusted. The intention in the code was to be able to regulate the volume of air, to any area or space, on an as-needed basis. If the volume dampers are installed but the adjusters are not accessible, this cannot be accomplished. The change to this section in the 2006 IMC will make access to the volume damper adjusters mandatory.

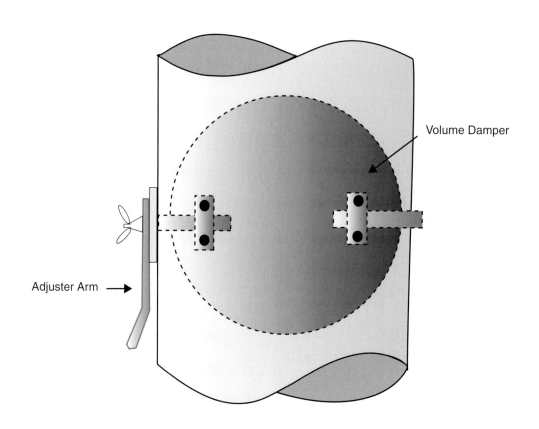

Volume Damper

Adjuster Arm

CHANGE TYPE.   Addition

CHANGE SUMMARY.   The addition of Section 603.17.2 will prohibit grilles, registers, and diffusers in duct systems from being located in floors in toilet rooms and bathing rooms when such floors are required by the IBC to have smooth, hard, nonabsorbent surfaces.

2006 CODE: **603.17.2 Prohibited Locations.**   Diffusers, registers, and grilles shall be prohibited in the floor or its upward extension within toilet and bathing room floors required by the *International Building Code* to have smooth, hard, nonabsorbent surfaces.

CHANGE SIGNIFICANCE.   This new addition to the 2006 IMC was submitted as a change to Section 1210.1 in the IBC and was added to the 2006 IMC. This change will prohibit grilles, registers, and diffusers in the floors and covering extension in commercial toilet rooms and bathrooms when the IBC requires the floor to have a smooth, hard, nonabsorbent surface.

When registers, grilles, or diffusers are located in the floor of a commercial bathroom or toilet room, water can get into the duct system and cause mold or mildew. This could occur from overflowing plumbing fixtures or the cleaning of the bathroom or toilet room floors. This new section has exempted dwelling units.

603.17.2
Prohibited Locations

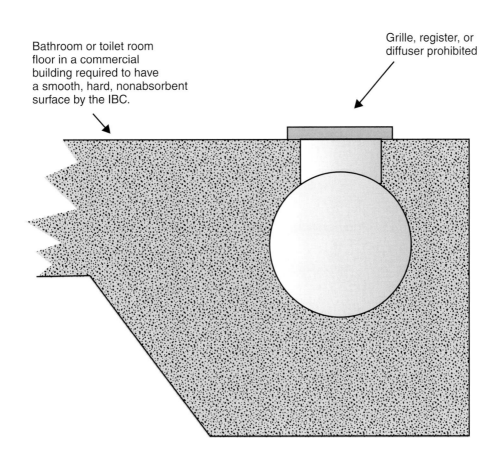

Bathroom or toilet room floor in a commercial building required to have a smooth, hard, nonabsorbent surface by the IBC.

Grille, register, or diffuser prohibited

607.5.5

Shaft Enclosures

CHANGE TYPE. Addition

CHANGE SUMMARY. The addition to Exception 2 in Section 607.5.5 in the IMC and Section 716.5.3 in the 2006 IBC will allow Group B and R occupancies that are equipped throughout with an automatic sprinkler system to have kitchen, clothes dryer, bathroom, and toilet room exhaust ducts subducted into a shaft without a smoke damper.

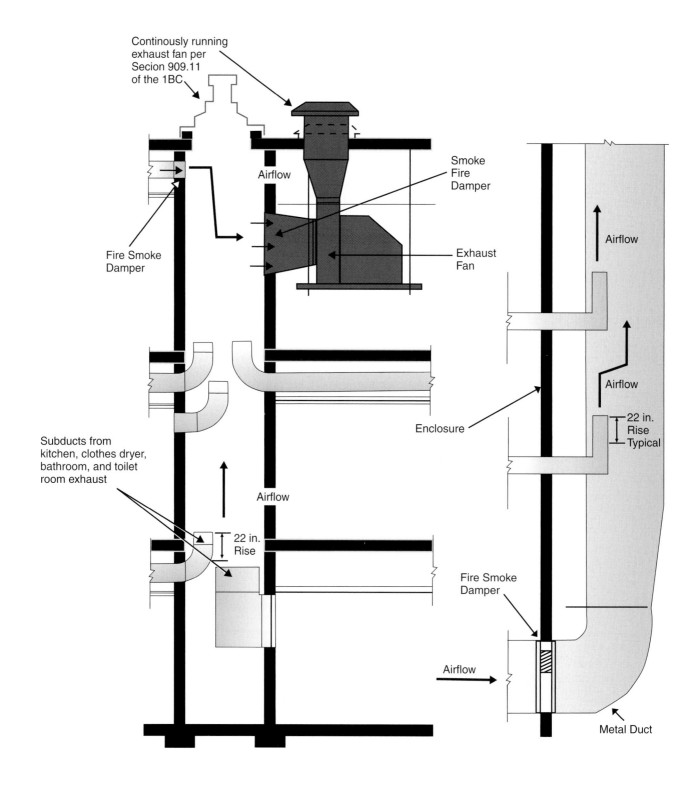

Continuously running exhaust fan per Secion 909.11 of the 1BC

Fire Smoke Damper

Subducts from kitchen, clothes dryer, bathroom, and toilet room exhaust

Airflow

Airflow

22 in. Rise

Smoke Fire Damper

Exhaust Fan

Airflow

Airflow

Airflow

Enclosure

22 in. Rise Typical

Fire Smoke Damper

Airflow

Metal Duct

2006 CODE: 607.5.5 Shaft Enclosures. ~~Ducts and air transfer openings shall not penetrate a shaft serving as an exit enclosure except as permitted by Section 1019.1.2 of the *International Building Code.*~~

~~**607.5.5.1 Penetrations of Shaft Enclosures.**~~ Shaft enclosures that are permitted to be penetrated by ducts and air transfer openings shall be protected with ~~approved~~ <u>listed</u> fire and smoke dampers installed in accordance with their listing.

Exceptions:
1. Fire dampers are not required at penetrations of shafts where:
    1.1. Steel exhaust subducts extend at least 22 inches (559 mm) vertically in exhaust shafts provided there is a continuous airflow upward to the ~~outside;~~ <u>outdoors;</u> ~~or~~
    1.2. Penetrations are tested in accordance with ASTM E 119 as part of the fire-resistance-rated assembly;
    1.3. Ducts are used as part of an approved smoke control system ~~designed and installed~~ in accordance with Section 909 of the *International Building Code,* and where the fire damper will interfere with the operation of the smoke control system; or
    1.4. The penetrations are in parking garage exhaust or supply shafts that are separated from other building shafts by not less than 2-hour fire-resistance-rated construction.

2. In Group B <u>and R</u> occupancies, equipped throughout with an automatic sprinkler system in accordance with Section 903.3.1.1 of the *International Building Code,* smoke dampers are not required at penetrations of shafts where~~:~~
    ~~**2.1.**~~ <u>Kitchen, clothes dryer,</u> ~~Bathroom~~ bathroom, and toilet room exhaust openings with steel exhaust subducts, having a wall thickness of at least 0.019 inch (0.48 mm) extend at least 22 inches (559 mm) vertically and the exhaust fan at the upper terminus is powered continuously in accordance with the provisions of Section 909.11 of the *International Building Code,* and maintains airflow upward to the ~~outside, or~~ <u>outdoors.</u>

    ~~**2.2.** Ducts are used as part of an approved smoke control system designed and installed in accordance with Section 909 of the *International Building Code,* and where the smoke damper will interfere with the operation of the smoke control system.~~

3. Smoke dampers are not required at penetration of exhaust or supply shafts in parking garages that are separated from other building shafts by not less than 2-hour fire-resistance-rated construction.

4. <u>Smoke dampers are not required at penetrations of shafts where ducts are used as part of an approved mechanical smoke control system designed in accordance with Section 909 of the *International Building Code* and where the smoke damper will interfere with the operation of the smoke control system.</u>

607.5.5 continues

607.5.5 continued

CHANGE SIGNIFICANCE. The change to this section in the 2006 IMC will now allow Group R occupancies to have subducted exhaust ducts from kitchen, clothes dryer, bathroom, and toilet room exhaust systems into a shaft enclosure without a smoke damper. In the previous codes there were exceptions to shaft penetrations that allowed for the elimination of smoke/fire dampers when exhaust ducts were extended at least 22 inches vertically into the shaft, but the elimination of the smoke damper applied only to Group B occupancies. The 2006 codes have added Group R occupancies to this section. This will now allow kitchen, clothes dryer, bathroom, and toilet room exhaust ducts to penetrate shaft enclosures without installation of a smoke or fire damper. This can be done only if the building has an automatic sprinkler system throughout that complies with Section 903.3.1.1 of the IBC and there is an exhaust fan at the upper terminus of the shaft that is powered continuously, in accordance with Section 909.11 of the IBC. This will be a tremendous savings in high-rise apartments and condominium buildings where this is normally done. It should be noted that where this section mentions kitchen exhaust it is referring to domestic kitchen exhaust systems and not commercial kitchen exhaust systems. It should also be noted that Section 505.1 in the 2006 IMC requires domestic range hoods to be ducted with sheet metal ducts. If a domestic kitchen exhaust duct was subducted into a shaft it would have to be ducted as shown in the drawing, or the shaft walls would have to be lined with sheet metal. The same is true for clothes dryers, per Section 504.6 in the 2006 IMC.

CHANGE TYPE

CHANGE SUM
calculate a spe
previous codes

2006 CODE:
systems in amn
at the emergen
Section 1105.6.

CHANGE TYPE.   Addition

CHANGE SUMMARY.   This is a new section in the 2006 IMC that refers to the International Fire Code (IFC; 606.10). This section will require refrigeration systems that are using flammable, toxic or highly toxic refrigerant or ammonia, with more than 6.6 pounds of refrigerant in the system, to have an emergency pressure control system.

2006 CODE: <u>**1105.9 Emergency Pressure Control System.** Refrigeration systems containing more than 6.6 pounds (3 kg) of flammable, toxic, or highly toxic refrigerant or ammonia shall be provided with an emergency pressure control system in accordance with Section 606.10 of the *International Fire Code.*</u>

CHANGE SIGNIFICANCE.   This new section that refers to the IFC will require a pressure relief system that will automatically relieve building pressure in a refrigeration system that has more than 6.6 pounds of a flammable, toxic, or highly toxic refrigerant or ammonia. This will be accomplished by installing an automatic pressure relief device that will allow building pressure in one part of the system to be relieved

1105.9 continues

1105.9

Emergency Pressure Control System

The lowest pressue zone in a refrigeration system shall be provided with a dedicated means of determining a rise in system pressure to within 15 psi of the set point for emergency pressure-relief devices. Activation of the over-pressure sensing device shall cause all compressors on the affected system to immediately stop.

High or Intermediate Pressure Zone

Piping to System

Low pressure Zone

Automatic crossover valve shall relieve pressure from the high to low side of the system if the pressure in the high or intermediate zone rises to within 15 psi of the set point for emergency pressure relief devices. This valve may be operated manually. This valve shall shut down all of the compressors in the zone it is connected to.

1106.3 continued

Exceptions:

1. Machinery rooms equipped with a vapor detector that will automatically start the ventilation system <u>at the emergency rate determined in accordance with Section 1105.6.4,</u> and <u>that will</u> actuate an alarm at a detection level not to exceed 1000 ppm; or

2. Machinery rooms conforming to the Class 1, Division 2, hazardous location classification requirements of ~~NPFA 70~~ <u>the</u> <u>ICC *Electrical Code*</u>.

CHANGE SIGNIFICANCE. This addition to Section 1106.3 will specify an actual method and rate for ventilation in an ammonia machinery room. It will also specify the ventilation rate if the ventilation system will be activated by a leak detector. Because of the toxic and flammable nature of ammonia, it was determined that the emergency ventilation rates specified in Section 1105.6.4 would be necessary.

CHANGE TYPE.   Addition

CHANGE SUMMARY.   The exception that was added to this section will not require electrical equipment in a machinery room, other than compressors, to be connected to an emergency shut-off switch when the system is using a nonflammable refrigerant.

2006 CODE: [F] 1106.5.1 Refrigeration System.   A clearly identified switch of the break-glass type shall provide off-only control of ~~all~~ electrically energized equipment and appliances in the machinery room, other than refrigerant leak detectors and machinery room ventilation.

>   **Exception:** In machinery rooms where only nonflammable refrigerants are used, electrical equipment and appliances, other than compressors, are not required to be provided with a cut-off switch.

CHANGE SIGNIFICANCE.   This exception to Section 1106.5.1 in the 2006 IMC will allow electrical equipment and appliances, other than the refrigeration compressors, to not be connected to an emergency shut-off switch when a nonflammable refrigerant is used in the sys-

1106.5.1 continues

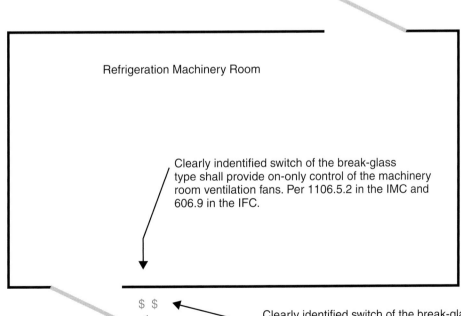

Refrigeration Machinery Room

Clearly indentified switch of the break-glass type shall provide on-only control of the machinery room ventilation fans. Per 1106.5.2 in the IMC and 606.9 in the IFC.

In machinery rooms where only nonflammable refrigerants are used, electrical equipment and appliances, other than compressors, are not required to be provided with a cutoff switch.

Clearly identified switch of the break-glass type shall provide off-only control of electrically energized equipment and appliances in the machinery room, other than refrigerant leak detectors and machinery room ventilation. Per Section 606.9 in the IFC, this switch shall be located at an approved location immediately outside the machinery room and adjacent to its principal entrance.

1106.5.1 continued

tem. This change was submitted to the *International Fire Code* and was added to the IMC. Where refrigerants are nonflammable, electrically energized equipment will not cause safety concerns. The intent of this section is to provide a safe environment for emergency personnel responding to an incident in a refrigeration machinery room. Because flammability is not an issue with nonflammable refrigerants, shutdown of electrical devices does not factor into the safety for emergency response personnel. It should be noted that this section conflicts with the ASHRAE 15–2004 Standard. Section 1101.6 in the IMC states that refrigeration systems shall comply with ASHRAE 15. Section 102.8 in the IMC states that where differences occur between provisions in the code and the reference standards, the provisions of the code shall apply. In this case, the new exception to Section 1106.5.1 would take precedence over the ASHRAE 15 Standard.

CHANGE TYPE.   Addition

CHANGE SUMMARY.   The new section 1301.5 will require all exterior above-grade fill piping to be removed when a fuel oil tank is abandoned or removed.

2006 CODE: <u>1301.5 Tanks Abandoned or Removed.</u>   <u>All exterior above-grade fill piping shall be removed when tanks are abandoned or removed. Tank abandonment and removal shall be in accordance with Section 3404.2.13 of the *International Fire Code.*</u>

CHANGE SIGNIFICANCE.   Section 3402.13 of the International Fire Code provides the requirements for abandoning or removing fuel oil tanks. It does not, however, require the removal of the above-grade fill piping associated with the tanks. Exterior fill and vent piping has to be removed because of the potential danger that hundreds of gallons of fuel oil could be accidentally delivered to the wrong address. There have been several instances of accidental filling where tanks have been removed but the fill pipe has remained. The oil delivery service sees only the exterior connection and may not be aware that the tank has been removed or disconnected. Piping must be removed, not just capped off; accidental filling has occurred when piping systems have

1301.5 continues

1301.5

Tanks Abandoned or Removed

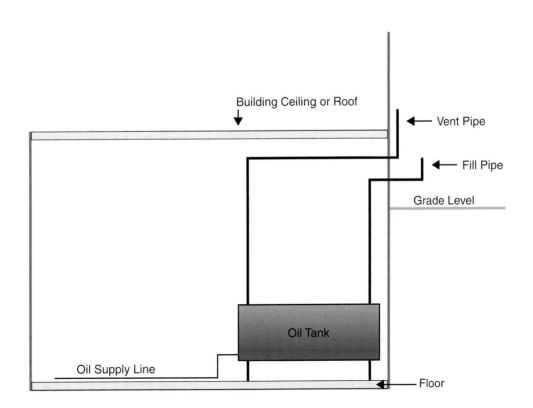

The above drawing shows a typical fuel oil storage system with the tank located in the basement and the vent and filler pipe on the outside of an exterior wall.

1301.5 continued

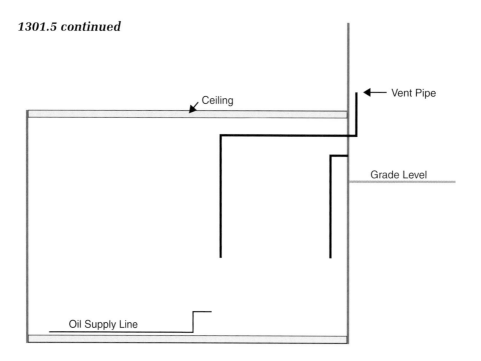

The above drawing shows the oil storage tank in the basement removed.
In this case code requires the above grade filler pipe to be removed. It
is a good idea, even though not required by code, to also remove the
above ground vent pipe.

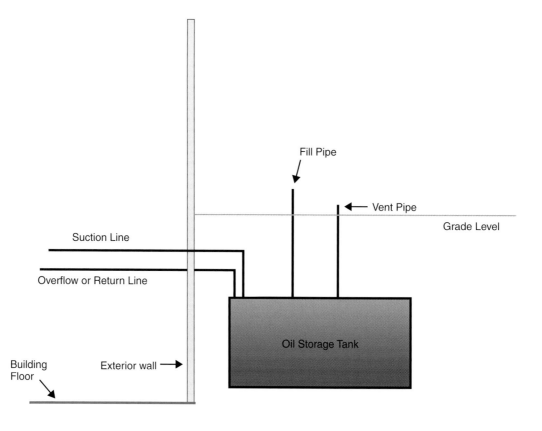

The above drawing shows a typical fuel oil storage system with the
tank located below grade outside of the building.

been simply capped off. When fuel oil contamination occurs, the cost of repairs and cleanup is extremely high, and some occurrences have led to condemnation of such structures.

Although it is not mentioned or required by this section in the code, it would be a good idea to remove any above-ground vent piping that was connected to the oil tank. This would prevent the vent pipe from being mistaken for a filler pipe and having the same problem as mentioned for the filler piping.

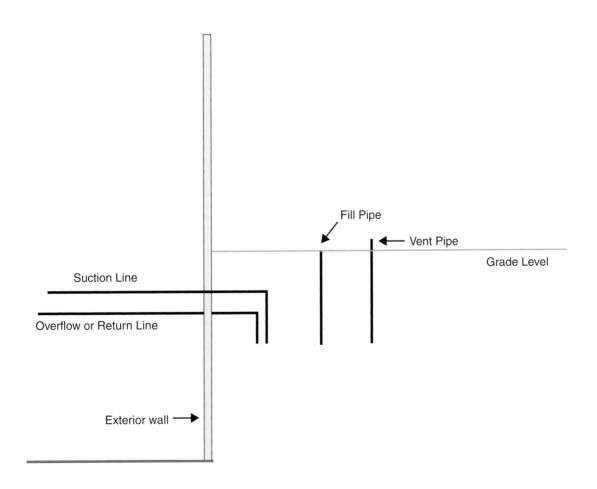

The above drawing shows the fuel oil storage tank, located below grade outside of the building, removed. The above grade fill pipe in this case would have to be removed. Even though code does not require the above grade vent pipe to be removed, it would be a good idea to remove it.

PART 3

International Fuel Gas Code

Chapters 1 Through 8

- **Chapter 1**   Administration No changes addressed
- **Chapter 2**   Definitions
- **Chapter 3**   General Regulations
- **Chapter 4**   Gas Piping Installations
- **Chapter 5**   Chimneys and Vents
- **Chapter 6**   Specific Appliances No changes addressed
- **Chapter 7**   Gaseous Hydrogen Systems
- **Chapter 8**   Referenced Standards No changes addressed

The *International Fuel Gas Code*® applies to the installation of fuel gas piping systems, fuel gas utilization equipment, gaseous hydrogen systems, and related accessories. Chapter 1 provides for the administration and enforcement of the code, assigning responsibility and authority to the code official. Chapter 2 contains definitions of terms specific to their use throughout the code. The general requirement provisions of Chapter 3 govern the approval and installation of all equipment and appliances regulated by the code. Requirements for the design and installation of gas piping systems are set out in Chapter 4 and include provisions for materials, components, fabrication, testing, inspection, operation, and maintenance of such systems. The scope of Chapter 5 includes factory-built chimneys, liners, vents, connectors, and masonry chimneys serving gas-fired appliances. Reference is made to the *International Mechanical Code* for chimneys serving appliances using other fuels and to the *International Building Code* for the construction requirements of masonry chimneys. Approval, design, and installation of specific appliances such as furnaces, boilers, water heaters, fireplaces, decorative appliances, room heaters, and clothes dryers are covered in Chapter 6. A recent addition to the code, Chapter 7, covers the developing technology of gaseous hydrogen systems, including hydrogen generation and refueling operations, and provides reference to the applicable provisions of the *International Fire Code*. Chapter 8 provides a complete list of standards referenced in various sections of the code. ∎

202

Definitions for Chimney Vent Connector and Appliance Fuel Connector

CHANGE TYPE. Addition

CHANGE SUMMARY. The definition for connector has been clarified to specify connection of the appliance to the chimney or vent. A new definition has been added for the appliance fuel connector to distinguish it from the chimney or vent connector.

2006 CODE: Section 202 General Definitions

CONNECTOR, CHIMNEY OR VENT. The pipe that connects an ~~approved~~ appliance to a chimney, ~~flue~~ or vent.

CONNECTOR, APPLIANCE (fuel). Rigid metallic pipe and fittings, semi-rigid metallic tubing and fittings or a listed and labeled device that connects an appliance to the gas piping system.

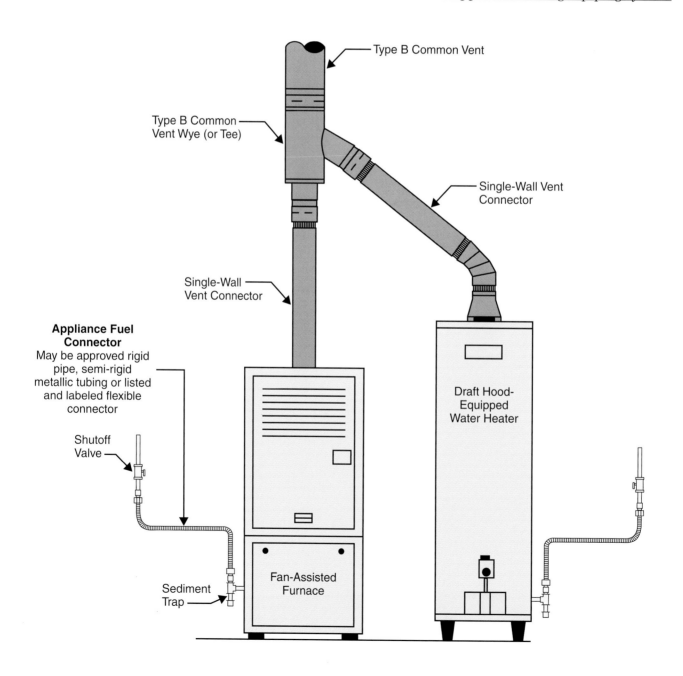

Type B Common Vent

Type B Common Vent Wye (or Tee)

Single-Wall Vent Connector

Single-Wall Vent Connector

Appliance Fuel Connector
May be approved rigid pipe, semi-rigid metallic tubing or listed and labeled flexible connector

Shutoff Valve

Sediment Trap

Fan-Assisted Furnace

Draft Hood-Equipped Water Heater

CHANGE SIGNIFICANCE. The term "connector" is used throughout the code to describe two different applications, connection of the appliance flue outlet to the flue gas exhaust system, and connection of the appliance to the fuel gas piping system. For example, Chapter 4, particularly Section 411, includes requirements for connectors between the appliance and the fuel gas piping system. The term *connector* in this use has previously been undefined, creating the possibility of confusion with the term as used to describe the connection of the appliance flue outlet to the vent. The revision to the current definition of chimney or vent connector and the addition of a definition for an appliance fuel connector will now clearly distinguish the two as separate items.

301.3

Listed and Labeled Appliances

CHANGE TYPE. Clarification

CHANGE SUMMARY. The intent of the code has been clarified to provide that gas-fired appliances must be installed and used in a manner that complies with the listing and labeling of the appliance.

2006 CODE: 301.3 Listed and Labeled. Appliances regulated by this code shall be listed and labeled <u>for the application in which they are used</u> unless otherwise approved in accordance with Section 105. The approval of unlisted appliances in accordance with Section 105 shall be based upon approved engineering evaluation.

CHANGE SIGNIFICANCE. Listing and labeling of gas-fired appliances by qualified, nationally recognized third-party agencies give assurance to the code official that an appliance will function for the intended purpose and operate safely. This section now mandates that, for listed and labeled appliances, the appliance must be installed and used in a manner consistent with the listing. The new language intends to prevent the installation of an appliance in an application or a location for which it has not been tested and approved. For example, furnaces that are listed for indoor residential use have occasionally been installed in outdoor locations serving commercial buildings.

In those uncommon instances where an appliance is not listed, the code has maintained the option for approval by the code official under Section 105, based on an engineering evaluation and testing by an approved testing agency. Unlisted appliances are often unique specialty equipment of very limited production or application. It is not feasible to obtain a listing for individual appliances. In these instances, an engineering evaluation gives adequate assurance to the code official that the appliance will perform as intended.

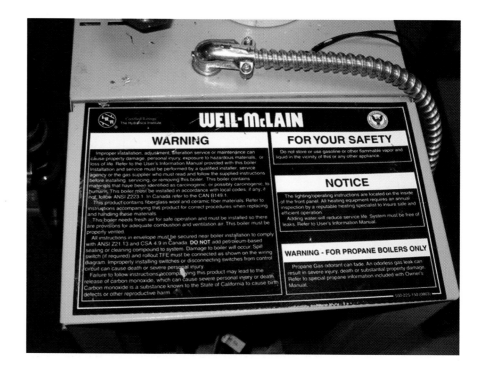

301.7

Appliance Fuel Types

CHANGE TYPE.   Modification

CHANGE SUMMARY.   Conversion of an appliance for connection to a different fuel type now requires that complete instructions for conversion must be provided by the serving gas supplier or the appliance manufacturer and must be included in the installation instructions. References to the altitude of the installation have been deleted.

2006 CODE: ~~301.7 Fuel Types.~~   ~~Appliances shall be designed for use with the type of fuel gas to which they will be connected and the altitude at which they are installed. Appliances that comprise parts of the installation shall not be converted for the usage of a different fuel, except where approved and converted in accordance with the manufacturer's instructions. The fuel gas input rate shall not be increased or decreased beyond the limit rating for the altitude at which the appliance is installed.~~

301.7 Fuel Types.   <u>Appliances shall be designed for use with the type of fuel gas that will be supplied to them.</u>

301.7 continues

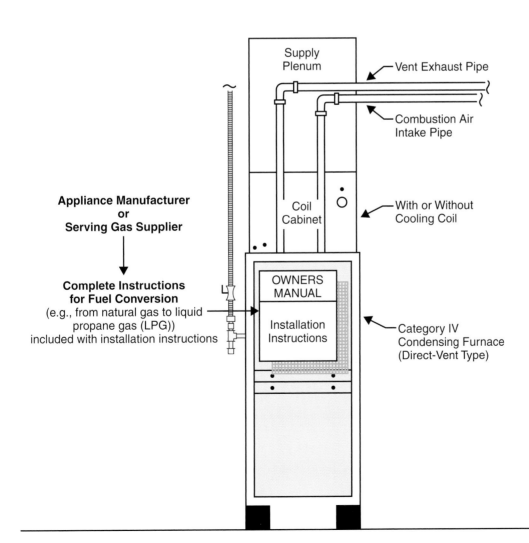

301.7 continued   <u>**301.7.1 Appliance Fuel Conversion.** Appliances shall not be converted to utilize a different fuel gas except where complete instructions for such conversion are provided in the installation instructions, by the serving gas supplier or by the appliance manufacturer.</u>

CHANGE SIGNIFICANCE. The code requires that appliances be connected to the type of fuel gas for which they are designed. Previously, conversion to a different type of fuel gas required only that the conversion was approved and accomplished in accordance with the manufacturer's instructions. The code now requires the means for conversion be specifically included in the installation instructions of the appliance and that such conversion instructions shall be provided by the serving gas supplier or the manufacturer.

Appliances are not specifically listed for altitude, and the text related to design for the altitude of the installation has been deleted. Adjustments to the appliance related to high altitude are made in the field at the time of installation.

CHANGE TYPE.   Clarification

CHANGE SUMMARY.   This section has been rewritten to clarify the five specific approved installations for gas-fired appliances in otherwise prohibited locations. Ambiguous language regarding the source of combustion air has also been deleted.

2006 CODE: 303.3 Prohibited Locations.   Appliances shall not be located in, ~~or obtain combustion air from, any of the following rooms or spaces:~~ sleeping rooms, bathrooms, toilet rooms, storage closets, or surgical rooms, or in a space that opens only into such rooms or spaces, except where the installation complies with one of the following:

1. ~~Sleeping rooms.~~
2. ~~Bathrooms.~~
3. ~~Toilet rooms.~~
4. ~~Storage closets.~~
5. ~~Surgical rooms.~~

~~Exceptions:~~

1. The appliance is a direct-vent appliance~~s that obtain all combustion air directly from the outdoors~~ installed in accordance with the conditions of the listing and the manufacturer's instructions.

2. Vented room heaters, wall furnaces, vented decorative appliances, vented gas fireplaces, vented gas fireplace heaters, and decorative appliances for installation in vented solid fuel-burning fireplaces, ~~provided that the room~~ are installed in rooms that meet~~s~~ the required volume criteria of Section 304.5.

3. ~~A single wall-mounted unvented room heater equipped with an oxygen depletion safety shutoff system and installed in a bathroom, provided that the input rating does not exceed 6000 Btu/h (1.76kW) and the bathroom meets the required volume criteria of Section 304.5.~~ A single wall-mounted unvented room heater is installed in a bathroom and such unvented room heater is equipped as specified in Section 621.6 and has an input rating not greater than 6000 Btu (1.76 kW). The bathroom shall meet the required volume criteria of Section 304.5.

4. ~~A single wall-mounted unvented room heater equipped with an oxygen depletion safety shutoff system and installed in a bedroom, provided that the input rating does not exceed 10,000 Btu/h (2.93 kW) and the bedroom meets the required volume criteria of Section 304.5.~~ A single wall-mounted unvented room heater is installed in a bedroom and such unvented room heater is equipped as specified in Section 621.6 and has an input rating not greater than 10,000 Btu (2.93 kW). The bedroom shall meet the required volume criteria of Section 304.5.

303.3
Appliance Prohibited Locations

303.3 continues

303.3 continued

5. ~~Appliances installed in an enclosure in which all combustion air is taken from the outdoors, in accordance with Section 304.6. Access to such enclosure shall be through a solid weather-stripped door, equipped with an approved self-closing device.~~ <u>The appliance is installed in a room or space that opens only into a bedroom or bathroom, such room or space is used for no other purpose, and is provided with a solid weather-stripped door equipped with an approved self-closing device. All combustion air shall be taken directly from the outdoors in accordance with Section 304.6.</u>

CHANGE SIGNIFICANCE.   The laundry list of five rooms prohibited from installations of gas-fired appliances has been reformatted into preferred code text, and the five unique exceptions have been rewritten as specific approved installations in such rooms. The effect of the change will be a clearer understanding of the requirements and more uniform enforcement by the code official. The requirement that the appliance shall not obtain combustion air from any of the prohibited

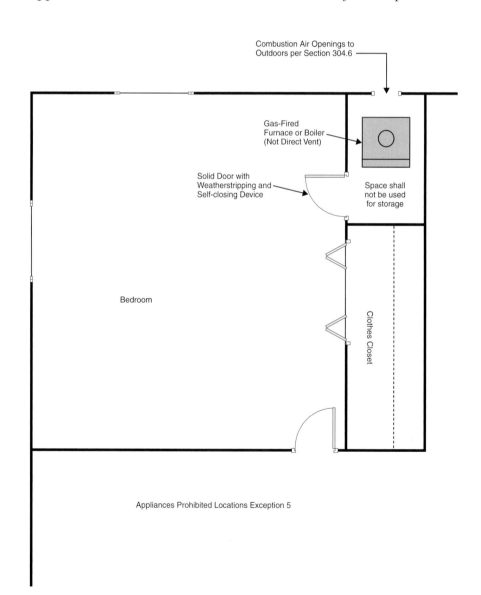

Appliances Prohibited Locations Exception 5

locations has been deleted. It was unclear as to the intent of the provision or the means to determine the source of combustion air (for example, small amounts of air infiltrating from communicating rooms).

In Item 1, the approved installation of a direct-vent appliance in otherwise prohibited locations has been reworded to delete mention of combustion air and instead specifies installation in accordance with the conditions of the listing and the manufacturer's instructions. By definition, a direct-vent appliance must obtain all combustion air from the outside atmosphere.

In Item 2, vented gas fireplaces and vented gas fireplace heaters have been added to the list of relatively low input rating appliances allowed, provided the room has sufficient volume for indoor combustion air, in accordance with section 304.5.

Items 3 and 4, allowing unvented appliances of limited input rating in bathrooms and bedrooms, respectively, have not changed appreciably other than replacement of the phrase "equipped with an oxygen depletion safety shutoff system" with a reference to the requirements of such a system in Section 621.6.

Item 5, allowing an appliance in an enclosed space with outside combustion air and a solid weather-stripped door equipped with an approved self-closing device, is now limited to a space used for no other purpose and which opens only to a bedroom or a bathroom.

303.4

Protection from Vehicle Impact Damage

CHANGE TYPE.   Modification

CHANGE SUMMARY.   Changes to this section clarify that appliances must be specifically protected from vehicle impact damage and that the means to provide protection is not limited to physical barriers.

2006 CODE: 303.4 Protection from ~~Physical~~ Vehicle Impact Damage.   Appliances shall not be installed in a location ~~where~~ subject to ~~physical~~ vehicle impact damage ~~unless~~ except where protected by an approved means ~~barriers meeting the requirements of the International Fire Code.~~

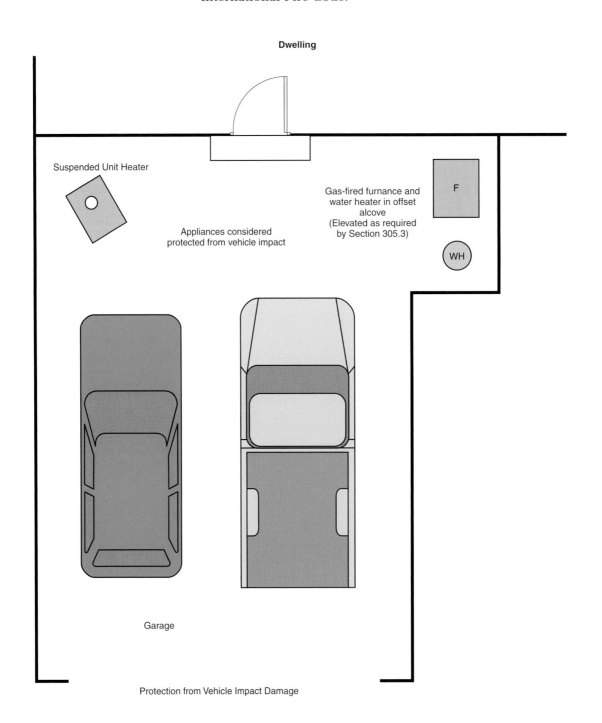

Protection from Vehicle Impact Damage

CHANGE SIGNIFICANCE. The intent of this section is to protect appliances from vehicle impact. Typically, these installations will be in garages, but the requirements could also apply to any location in the vicinity of vehicles. The term *physical damage* was considered too broad, and the more specific language will remove inconsistencies in interpretation of this section.

The term *barrier* has been replaced by *means* for protection. A physical barrier is only one way to protect appliances from vehicle impact. The code now recognizes other methods to achieve protection. For example, appliances may be positioned in offsets out of the path of vehicle movement, or they may be elevated above the height of the vehicle, such as in the installation of a suspended unit heater. The reference to the *International Fire Code* (IFC) has been deleted because it could lead to the incorrect conclusion that the intent was to protect from the storage of flammable and hazardous materials.

305.3

Elevation of Ignition Source in Garages and Hazardous Locations

CHANGE TYPE.   Addition

CHANGE SUMMARY.   The exception to Section 305.3 has been modified to reflect that only appliances specifically listed as flammable vapor ignition–resistant are exempt from the elevation requirements. Requirements of the *International Building Code* (IBC) for separating rooms containing gas-fired appliances from a parking garage are now included in this code. An exception to the separation provision has been added to allow the provisions of Section 305.4 related to appliance installation in a public garage.

2006 CODE: 305.3 Elevation of Ignition Source. Equipment and appliances having an ignition source shall be elevated such that the source of ignition is not less than 18 inches (457 mm) above the floor in hazardous locations and public garages, private garages, repair garages, motor fuel dispensing facilities and parking garages. For the purpose of this section, rooms or spaces that are not part of the living space of a dwelling unit and that communicate directly with a private garage through openings shall be considered to be part of the private garage.

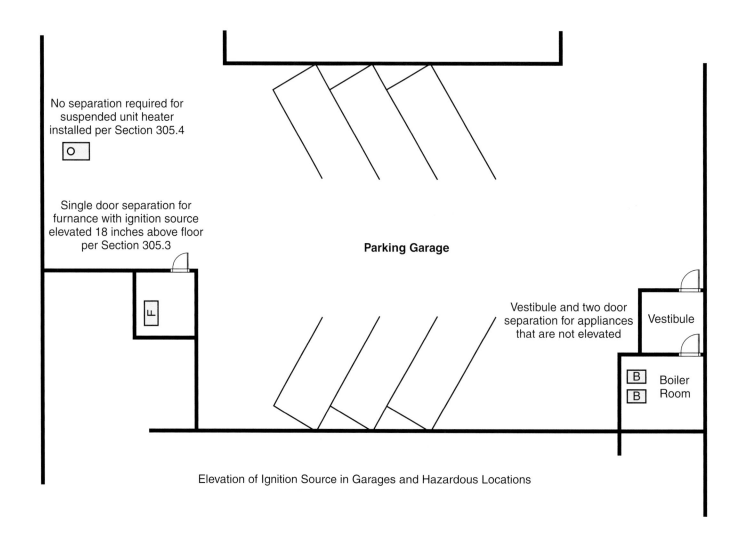

No separation required for suspended unit heater installed per Section 305.4

Single door separation for furnance with ignition source elevated 18 inches above floor per Section 305.3

Parking Garage

Vestibule and two door separation for appliances that are not elevated

Vestibule

Boiler Room

Elevation of Ignition Source in Garages and Hazardous Locations

Exception: Elevation of the ignition source is not required for appliances that are listed as flammable vapor <u>ignition</u> resistant ~~and for installation without elevation~~.

305.3.1 Parking Garages. <u>Connection of a parking garage with any room in which there is a fuel-fired appliance shall be by means of a vestibule providing a two-doorway separation, except that a single door is permitted where the sources of ignition in the appliance are elevated in accordance with Section 305.3.</u>

> **Exception:** <u>This section shall not apply to appliance installations complying with Section 305.4.</u>

CHANGE SIGNIFICANCE. The exception to Section 305.3 has been modified to delete the ambiguous phrase "and for installation without elevation" and clarifies that an appliance must be specifically listed as flammable vapor ignition–resistant. The new language reflects the correct terminology for a listed appliance that does not need to be elevated in garages and hazardous locations.

A new section has been added to include text from the IBC requiring separation from a parking garage. The provision mandates a single-door separation if the ignition source of the appliance is elevated 18 inches above the floor, in accordance with Section 305.3. A vestibule and two-doorway separation between the room containing the gas-fired appliance and the parking garage is required if the appliance is not elevated. Since this provision involves the installation of a gas-fired appliance, duplicating this text from the IBC prevents this requirement from being overlooked by the mechanical inspector.

An exception has been added to clarify that appliances elevated a minimum of 8 feet above the floor or at least 2 feet above vehicle height (where motor vehicles exceed 6 feet in height), in accordance with Section 305.4, do not require separation from the parking garage. Suspended unit heaters meeting the height clearance requirements satisfy the provisions for elevation in Section 305.3 and for protection from vehicle impact in Section 303.4. As such they are allowed to be installed in parking garages. Separation of spaces containing suspended unit heaters from parking garages would serve no purpose.

307.1

Evaporators and Cooling Coils

CHANGE TYPE. Addition

CHANGE SUMMARY. A reference to the *International Mechanical Code* (IMC) regarding drainage systems for evaporators and cooling coils has been added.

2006 CODE: <u>**307.1 Evaporators and Cooling Coils.** Condensate drainage systems shall be provided for equipment and appliances containing evaporators and cooling coils in accordance with the *International Mechanical Code.*</u>

CHANGE SIGNIFICANCE. The code only addresses condensate produced as a result of combustion. However, gas-fired appliances often also contain cooling coils which produce condensate. The reference to the IMC completes the information necessary for removal of condensate produced by combustion or cooling.

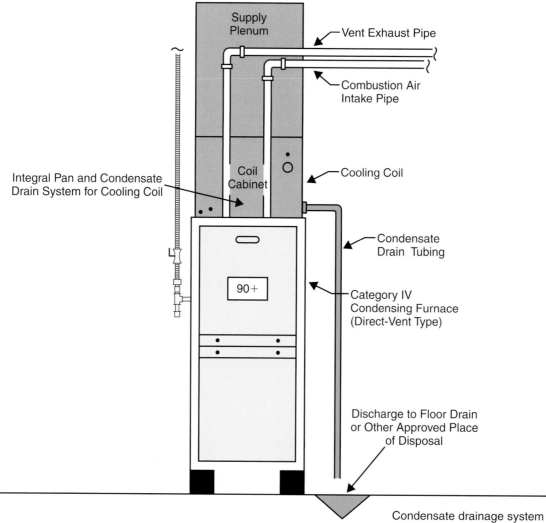

Evaporators and Cooling Coils

CHANGE TYPE. Addition

CHANGE SUMMARY. An auxiliary drain pan is now required to protect building components from water damage in the case of failure of the condensate drainage system of Category IV appliances. An exception allows an automatic appliance shut-down in lieu of the drain pan.

2006 CODE: **307.5 Auxiliary Drain Pan.** Category IV condensing appliances shall be provided with an auxiliary drain pan where dam-

307.5 continues

307.5
Auxiliary Drain Pan

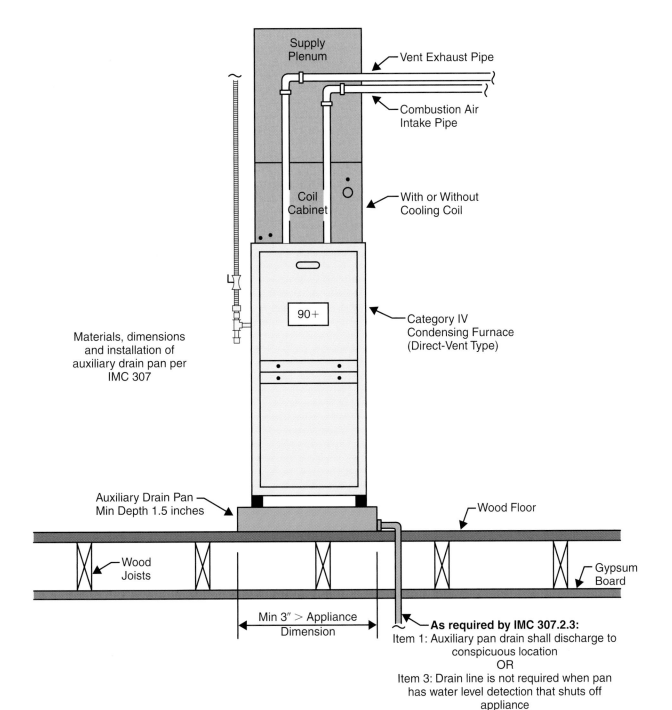

Supply Plenum

Vent Exhaust Pipe

Combustion Air Intake Pipe

Coil Cabinet

With or Without Cooling Coil

90+

Category IV Condensing Furnace (Direct-Vent Type)

Materials, dimensions and installation of auxiliary drain pan per IMC 307

Auxiliary Drain Pan Min Depth 1.5 inches

Wood Floor

Wood Joists

Gypsum Board

Min 3″ > Appliance Dimension

As required by IMC 307.2.3:
Item 1: Auxiliary pan drain shall discharge to conspicuous location
OR
Item 3: Drain line is not required when pan has water level detection that shuts off appliance

307.5 continued

age to any building component will occur as a result of stoppage in the condensate drainage system. Such pan shall be installed in accordance with the applicable provisions of Section 307 of the *International Mechanical Code.*

> **Exception:** An auxiliary drain pan shall not be required for appliances that automatically shut down operation in the event of a stoppage in the condensate drainage system.

CHANGE SIGNIFICANCE. The new provisions for a drain pan for those appliances producing condensate as a byproduct of combustion match similar requirements in the IMC for cooling coil condensate appliances. The auxiliary pan is required in case the condensate drainage system becomes plugged, but only where the resulting water spillage would cause damage to building components. This would apply, for example, to a Category IV condensing furnace installed on the second floor or in the attic of a dwelling where water leakage would damage the drywall ceiling of the living space below.

The drain pan must be installed in accordance with the requirements of the applicable portions of Section 307 of the IMC. As such, the auxiliary drain pan shall have a separate drain discharging to a conspicuous location; must be not less than 3 inches larger in dimensions than the unit served, with a minimum depth of 1.5 inches; and must be of corrosion-resistant material. Metallic pans shall have a minimum thickness of not less than 0.0276-inch (0.7 mm) galvanized sheet metal. Nonmetallic pans shall have a minimum thickness of not less than 0.0625 inch (1.6 mm).

The exception allows omission of the auxiliary drain pan if the appliance is designed to automatically shut down should the primary condensate drainage system stop functioning.

CHANGE TYPE. Modification

CHANGE SUMMARY. Gas piping serving a townhouse is not allowed to pass through any other townhouse.

2006 CODE: 404.1 Prohibited Locations. Piping shall not be installed in or through a circulating air duct, clothes chute, chimney or gas vent, ventilating duct, dumbwaiter or elevator shaft. <u>Piping installed downstream of the point of delivery shall not extend through any townhouse unit other than the unit served by such piping.</u>

CHANGE SIGNIFICANCE. Townhouse construction creates some unique concerns regarding the arrangement and routing of gas piping. The premise for the code change is that gas piping serving a dwelling should be under the control of the occupant of that unit. It is not uncommon in townhouse construction for the gas meters serving all of the units to be grouped in one location. Gas piping is then routed through attics or crawl spaces of adjoining units to reach the unit being served. The change will necessitate routing of gas piping from the meter location underground, outside of the building, to serve each unit, or disbursing the meters in such a way to serve each unit without gas piping passing through other units.

404.1

Prohibited Locations for Gas Piping

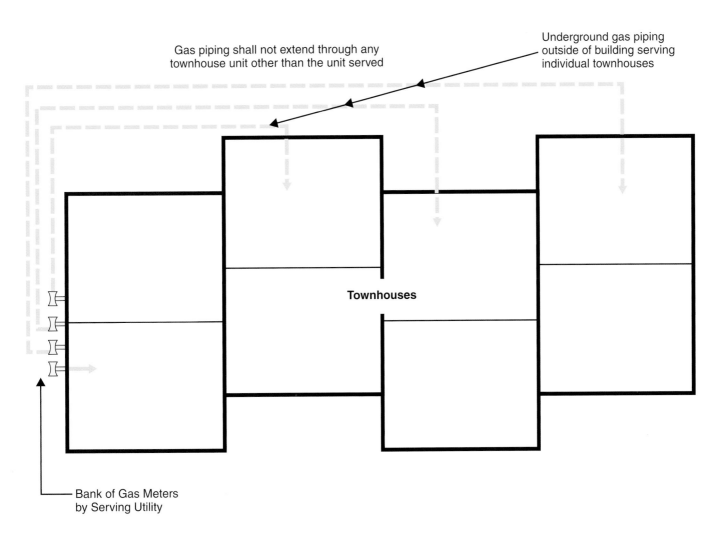

Gas piping shall not extend through any townhouse unit other than the unit served

Underground gas piping outside of building serving individual townhouses

Townhouses

Bank of Gas Meters by Serving Utility

404.5, 704.1.2.3.5

Protection Against Physical Damage

CHANGE TYPE. Modification

CHANGE SUMMARY. For gas piping and gaseous hydrogen system piping of materials other than steel installed in wood framing, the distance from the edge of the framing has been increased from 1 inch to 1.5 inches to provide protection from fastener penetration. Piping installed less than 1.5 inches from the face must be protected by steel shield plates.

2006 CODE: **404.5 Protection Against Physical Damage.** In concealed locations, where piping other than black or galvanized steel is installed through holes or notches in wood studs, joists, rafters, or similar members less than ~~1~~ 1.5 inches (~~25~~ 38 mm) from the nearest edge of the member, the pipe shall be protected by shield plates. Shield plates shall be a minimum of $\frac{1}{16}$-inch-thick (1.6 mm) steel, shall cover the area of the pipe where the member is notched or bored, and shall extend a minimum of 4 inches (102 mm) above sole plates, below top plates and to each side of a stud, joist or rafter.

704.1.2.3.5 Protection Against Physical Damage. In concealed locations, where piping other than stainless steel piping, stainless steel tubing or black steel is installed through holes or notches in wood studs, joists, rafters or similar members less than ~~1~~1.5 inches (~~25~~ 38 mm) from the nearest edge of the member, the pipe shall be protected by shield plates. Shield plates shall be a minimum of $\frac{1}{16}$-inch-thick (1.6 mm) steel, shall cover the area of the pipe where the member is notched or bored, and shall extend a minimum of 4 inches (102 mm) above sole plates, below top plates and to each side of a stud, joist or rafter.

CHANGE SIGNIFICANCE. Concealed fuel gas piping and gaseous hydrogen system piping subject to penetration by nails or screws being driven into wood framing members must be protected to prevent a potentially hazardous leak. Previously, protection by a steel shield plate was required if the piping (other than steel piping) was less than 1 inch from the edge of the framing member. The distance has been increased to 1.5 inches to provide a higher degree of protection from fastener penetration and to bring consistency with the mechanical and plumbing piping protection provisions of the *International Residential Code* (IRC), the IMC, and the *International Plumbing Code* (IPC). In a typical residential drywall application, a common $1\frac{5}{8}$–inch drywall screw attaching $\frac{1}{2}$ inch gypsum board would penetrate the framing member at least $1\frac{1}{8}$ inches plus countersinking depth. Similarly, nails from a pneumatic nailer attaching paneling or trim may penetrate the framing to a depth greater than 1 inch, depending on the length of the nail and depth of the countersink.

Black or galvanized steel fuel gas piping is resistant to penetration and requires no additional physical protection. Similarly, hydrogen system piping, which must be specifically approved and compatible with hydrogen gas, does not require additional protection if it is stainless steel piping, stainless steel tubing, or black steel. These provisions for protection of hydrogen system piping in concealed locations has little or no application since changes to Sections 704.1.2.3.1 and 704.1.2.3.2 do not allow the piping to be concealed by the surface of any wall, floor, or ceiling.

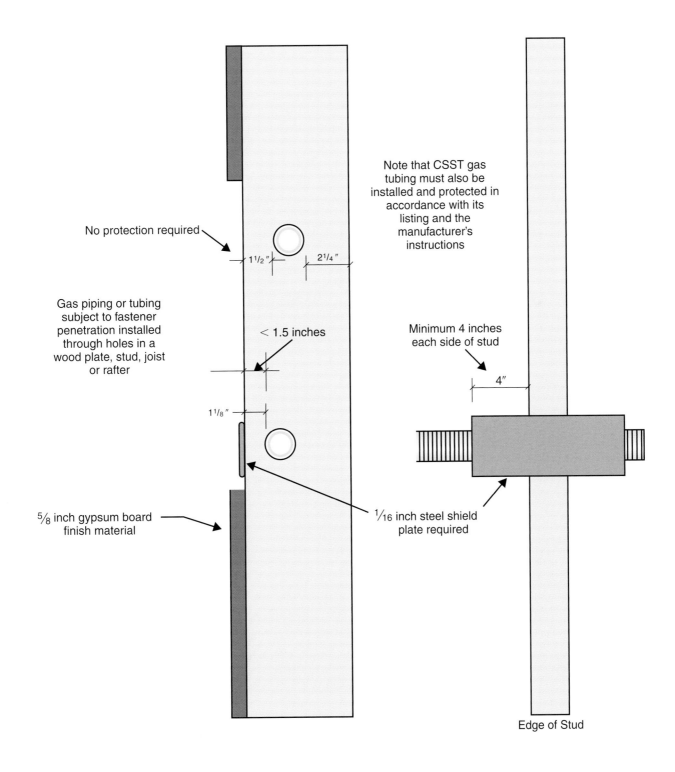

No protection required

1 1/2 "   2 1/4 "

Note that CSST gas tubing must also be installed and protected in accordance with its listing and the manufacturer's instructions

Gas piping or tubing subject to fastener penetration installed through holes in a wood plate, stud, joist or rafter

< 1.5 inches

Minimum 4 inches each side of stud

4"

1 1/8 "

5/8 inch gypsum board finish material

1/16 inch steel shield plate required

Edge of Stud

404.6

Conduit for Piping in Solid Floors

CHANGE TYPE.   Modification

CHANGE SUMMARY.   The word casing has been replaced by conduit as a more appropriate term for the protection of fuel gas piping installed in solid floors. A new requirement provides for venting of the conduit.

2006 CODE: 404.6 Piping in Solid Floors.   Piping in solid floors shall be laid in channels in the floor and covered in a manner that will allow access to the piping with a minimum amount of damage to the building. Where such piping is subject to exposure to excessive moisture or corrosive substances, the piping shall be protected in an approved manner. As an alternative to installation in channels, the piping shall be installed in a ~~casing~~ <u>conduit</u> of Schedule 40 steel, wrought iron, PVC or ABS pipe with tightly sealed ends and joints. Both ends of such ~~casing~~ <u>conduit</u> shall extend not less than 2 inches (51 mm) beyond the point where the pipe emerges from the floor. <u>The conduit shall be vented above grade to the outdoors and shall be installed so as to prevent the entry of water and insects.</u>

CHANGE SIGNIFICANCE.   The term *conduit* is more appropriate as a means to contain any leakage and protect gas piping installed in con-

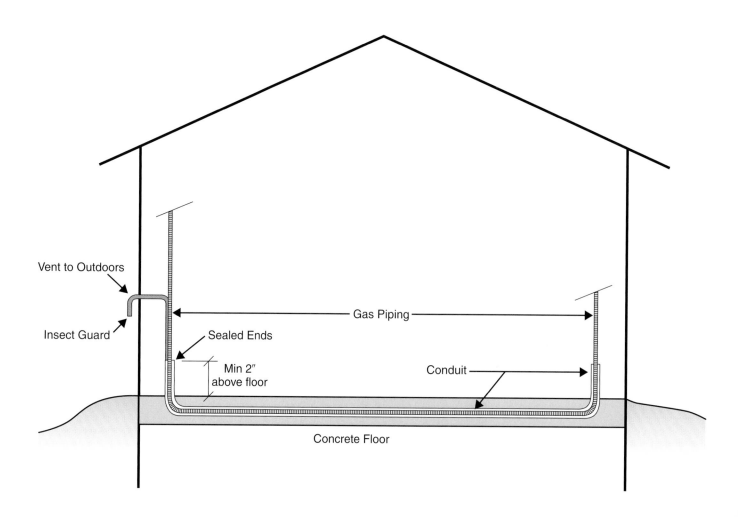

crete or other solid floors (as an alternative to laying the piping in channels). The previous term, *casing,* implies a structural requirement, which is not the intent. The new requirement mandates venting of the conduit above ground in case of a leak in the gas piping and further requires a means to prevent entry of insects or water into the conduit. Changes to this section are consistent with Section 404.11, which requires that gas piping installed underground beneath buildings be in a conduit vented above grade and be installed in such a manner to prevent the entrance of water and insects.

410.3

Venting of Regulators

CHANGE TYPE. Addition

CHANGE SUMMARY. Venting of regulators has been clarified to require venting directly to the outdoors and to prevent the entry of insects into the vent. A new section has been created to provide more detailed requirements on the size and installation of regulator vents and to limit when vents can be part of a manifold system.

2006 CODE: 410.3 Venting of Regulators. Pressure regulators that require a vent shall ~~have an independent vent to the outside of the building~~ be vented directly to the outdoors. The vent shall be designed to prevent the entry of insects, water ~~or~~ and foreign objects.

> **Exception:** A vent to the ~~outside of the building~~ outdoors is not required for regulators equipped with and labeled for utilization with an approved vent-limiting device~~s~~ installed in accordance with the manufacturer's instructions.

410.3.1 Vent Piping. Vent piping shall be not smaller than the vent connection on the pressure regulating device. Vent piping serving relief vents and combination relief and breather vents shall be run inde-

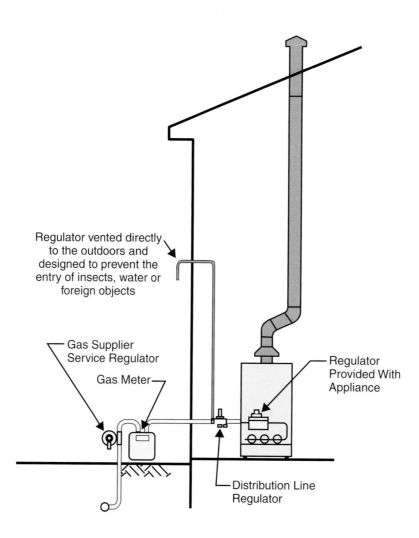

Regulator vented directly to the outdoors and designed to prevent the entry of insects, water or foreign objects

Gas Supplier Service Regulator

Gas Meter

Regulator Provided With Appliance

Distribution Line Regulator

pendently to the outdoors and shall serve only a single device vent. Vent piping serving only breather vents is permitted to be connected in a manifold arrangement where sized in accordance with an approved design that minimizes back pressure in the event of diaphragm rupture.

Section 202 General Definitions

VENT PIPING

Breather. Piping run from a pressure regulating device to the outdoors, designed to provide a reference to atmospheric pressure. If the device incorporates an integral pressure relief mechanism, a breather vent can also serve as a relief vent.

Relief. Piping run from a pressure-regulating or pressure-limiting device to the outdoors, designed to provide for the safe venting of gas in the event of excessive pressure in the gas piping system.

CHANGE SIGNIFICANCE.   Venting provisions for pressure regulators have been clarified to require that the regulator be vented directly to the outdoors. A new requirement has been added to prevent the vent from being blocked by the entry of insects at the vent termination.

The new section requires that the vent not be reduced to a smaller diameter than the vent connection on the regulator. It also clarifies the circumstances allowing a manifold arrangement of vents. Relief vents, which provide for venting of gas in the event of excessive pressure, are not allowed to be grouped in a manifold. Each vent line of a relief vent must be sized individually for the full release of gas. Breather vents for pressure regulators, designed only to connect to the outside atmosphere as a reference, may be arranged in a manifold, provided the vent piping is sized in accordance with engineering principles and the design is approved by the code official. The breather vent piping design must provide for the full release of gas should the diaphragm of one of the pressure regulator devices fail.

The new definitions for Breather and Relief Vent Piping are important to understanding the terminology used in the new Section 410.3.1.

411

Appliance Fuel Gas Connections

CHANGE TYPE. Modification

CHANGE SUMMARY. Section 411 has been entirely reorganized to clarify the requirements for fuel gas connectors. Methods for the connection of manufactured homes and outdoor appliances to the gas piping system are now recognized and specified. Corrugated Stainless Steel Tubing (CSST) is now specifically recognized and approved for connecting the appliance to the gas piping system. Commercial cooking appliances subject to moving are limited to connection with a listed and labeled connector in accordance with ANSI Z21.69.

2006 CODE:

Section 411 (IFGC) Appliance <u>and Manufactured Home</u> Connections

411.1 Connecting Appliances. <u>Except as required by Section 411.1.1,</u> appliances shall be connected to the piping system by one of the following:

1. Rigid metallic pipe and fittings.
2. <u>Corrugated Stainless Steel Tubing (CSST) where installed in accordance with the manufacturer's instructions.</u>
3. Semirigid metallic tubing and metallic fittings. Lengths shall not exceed 6 feet (1829 mm) and shall be located entirely in the same room as the appliance. Semirigid metallic tubing

Corrugated Stainless Steel Tubing (CSST) is permitted to connect appliances if installed in accordance with the manufacturer's instructions.

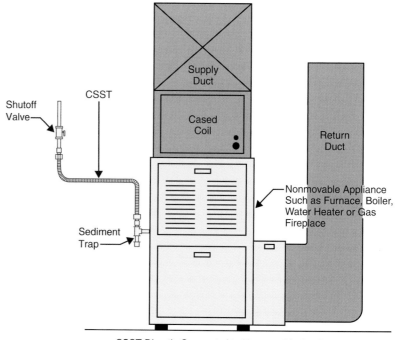

CSST Directly Connected to Nonmovable Appliance

Connecting Appliances to the Gas Piping System

shall not enter a motor-operated appliance through an unprotected knockout opening.

4. Listed and labeled appliance connectors <u>in compliance with ANSI Z21.24 and</u> installed in accordance with the manufacturer's installation instructions and located entirely in the same room as the appliance.

5. Listed and labeled quick-disconnect devices used in conjunction with listed and labeled appliance connectors.

6. Listed and labeled convenience outlets used in conjunction with listed and labeled appliance connectors.

7. Listed and labeled appliance connectors complying with ANSI Z21.69 and listed for use with food service equipment having casters, or that is otherwise subject to movement for cleaning, and other large movable equipment.

8. <u>Listed and labeled outdoor appliance connectors in compliance with ANSI Z21.75/CSA 6.27 and installed in accordance with the manufacturer's installation instructions.</u>

411.1.1 Commercial Cooking Appliances. <u>Commercial cooking appliances that are moved for cleaning and sanitation purposes shall be connected to the piping system with an appliance connector listed as complying with ANSI Z21.69.</u>

~~**411.1.2 Appliance Fuel Connectors.** Connectors shall have an overall length not to exceed 3 feet (914 mm), except for range and domestic clothes dryer connectors, which shall not exceed 6 feet (1829 mm) in length. Connectors shall not be concealed within, or extended through, walls, floors, partitions, ceilings or appliance housings. A shutoff valve not less than the nominal size of the connector shall be installed ahead of the connector in accordance with Section 409.5. Connectors shall be sized to provide the total demand of the connected appliance.~~

~~**Exception:** Fireplace inserts factory-equipped with grommets, sleeves, or other means of protection in accordance with the listing of the appliance.~~

411.1.2 Protection Against Damage. Connectors and tubing shall be installed so as to be protected from physical damage.

411.1.3 Connector Installation. <u>Appliance fuel connectors shall be installed in accordance with the manufacturer's instructions and Sections 411.1.3.1 through 411.1.3.4.</u>

411.1.3.1 Maximum Length. <u>Connectors shall have an overall length not to exceed 3 feet (914 mm), except for range and domestic clothes dryer connectors, which shall not exceed 6 feet (1829 mm) in overall length. Measurement shall be made along the centerline of the connector. Only one connector shall be used for each appliance.</u>

411 continues

411 continued

Exception: Rigid metallic piping used to connect an appliance to the piping system shall be permitted to have a total length greater than 3 feet (914 mm), provided that the connecting pipe is sized as part of the piping system in accordance with Section 402, and the location of the equipment shutoff valve complies with Section 409.5.

411.1.3.2 Minimum Size.   Connectors shall have the capacity for the total demand of the connected appliance.

411.1.3.3 Prohibited Locations and Penetrations.   Connectors shall not be concealed within, or extended through, walls, floors, partitions, ceilings, or appliance housings.

Exception: Fireplace inserts that are factory-equipped with grommets, sleeves, or other means of protection in accordance with the listing of the appliance.

411.1.3.4 Shutoff Valve.   A shutoff valve not less than the nominal size of the connector shall be installed ahead of the connector in accordance with Section 409.5.

411.2 Manufactured Home Connections.   Manufactured homes shall be connected to the distribution piping system by one of the following materials:

1. Metallic pipe in accordance with Section 403.4.
2. Metallic tubing in accordance with Section 403.5.
3. Listed and labeled connectors in compliance with ANSI Z21.75/CSA 6.27 and installed in accordance with the manufacturer's installation instructions.

CHANGE SIGNIFICANCE.   Section 411 has undergone extensive revision and reorganization. The charging statement in Section 411.1, recognizing a list of approved connectors, now directs the user to specific limitations in Section 411.1.1 regarding listed connectors required for any movable commercial cooking appliance. Commercial cooking appliances that are moved for cleaning must have connectors specifically listed in accordance with ANSI Z21.69. These connectors are reinforced and designed to withstand repeated movement without damage to the components.

CSST is now specifically recognized and approved for connecting the appliance to the gas piping system. Such use as a connector has been permitted in accordance with the manufacturer's instructions. Its inclusion in the text of the code will bring clarity and consistency to the enforcement of this installation.

The provision for listed and labeled appliance connectors now references the appropriate ANSI Z21.24 standard.

Listed and labeled outdoor appliance connectors in compliance with ANSI Z21.75/CSA 6.27 are now recognized in the text.

Section 411.1.3, previously a lengthy paragraph with multiple diverse requirements, has been broken down into a number of subsections, each dealing with a specific requirement. The resulting list of subsections provides a clearer understanding of the connector installation requirements and is compatible with ICC text style and format. The exception, which appeared to apply to the entire list of provisions in the paragraph, now correctly applies only to the applicable portion in Section 411.1.3.3.

In the charging statement of Section 411.1.3, wording has been added to emphasize that fuel gas connectors must be installed in accordance with the manufacturer's instructions, as well as the list of requirements that follow. In Section 411.1.3.1, the code now provides that the method of measurement for the maximum length of a connector shall be made along the centerline of the connector. Consequently, it is clear that the correct application of the code is determined by the actual length of the connector, not the relative location of the appliance in relation to its connection to the gas piping system. It is further noted that only one connector shall be used for each appliance, consistent with the requirements of ANSI Z223.1.

Manufactured homes are now included with appliances in the title of Section 411. A new Section 411.2, for manufactured home connections, has been added to recognize the use of listed and labeled fuel gas connectors in accordance with ANSI Z21.75/CSA 6.27–01, a standard that is now referenced in Chapter 8 of the code. The code also recognizes metallic tubing and rigid metallic pipe for connecting mobile homes to the gas piping system.

413.2.3

Residential Compressed Natural Gas Motor Vehicle Fuel Dispensing Appliances

CHANGE TYPE.   Addition

CHANGE SUMMARY.   New requirements have been added specific to residential applications of Compressed Natural Gas (CNG) Motor Vehicle Fuel Dispensing Appliances. Residential CNG fueling appliances must be listed and are of limited natural gas delivery capacity. The appliance may be installed in an indoor location, provided it is vented to the outdoors and a gas detector is installed near the ceiling.

2006 CODE: **413.2.3 General.** Residential fueling appliances shall be listed. The capacity of a residential fueling appliance shall not exceed 5 standard cubic feet per minute (0.14 standard cubic meter/min) of natural gas.

413.3 Location of Dispensing Operations and Equipment. Compression, storage, and dispensing equipment shall be located above ground outside.

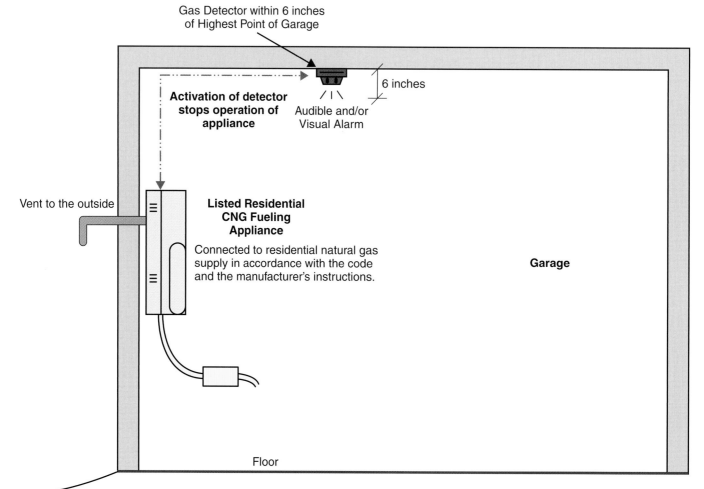

Gas Detector within 6 inches of Highest Point of Garage

6 inches

Activation of detector stops operation of appliance

Audible and/or Visual Alarm

Vent to the outside

Listed Residential CNG Fueling Appliance

Connected to residential natural gas supply in accordance with the code and the manufacturer's instructions.

Garage

Floor

Residential Compressed Natural Gas Motor Vehicle Fuel Dispensing Appliances

Exceptions:

1. Compression, storage or dispensing equipment is allowed in buildings of noncombustible construction, as set forth in the *International Building Code*, which are unenclosed for three-quarters or more of the perimeter.

2. Compression, storage and dispensing equipment is allowed to be located indoors in accordance with the *International Fire Code*.

3. Residential fueling appliances and equipment shall be allowed to be installed indoors in accordance with the equipment manufacturer's instructions and Section 413.4.3.

413.3.1 Location on Property.   In addition to the fuel-dispensing requirements of the *International Fire Code*, compression, storage and dispensing equipment other than residential fueling appliances shall not be installed:

1. **through 5.** (No change to current text)

413.4 Residential Fueling Appliance Installation. Residential fueling appliances shall be installed in accordance with Sections 413.4.1 through 413.4.3.

413.4.1 Gas Connections. Residential fueling appliances shall be connected to the premise's gas piping system without causing damage to the piping system or the connection to the internal appliance apparatus.

413.4.2 Outdoor Installation. Residential fueling appliances located outdoors shall be installed on a firm, noncombustible base.

413.4.3 Indoor Installation. Where located indoors, residential fueling appliances shall be vented to the outdoors. A gas detector set to operate at one-fifth of the lower limit of flammability of natural gas shall be installed in the room or space containing the appliance. The detector shall be located within 6 inches (150 mm) of the highest point in the room or space. The detector shall stop the operation of the appliance and activate an audible or a visual alarm.

CHANGE SIGNIFICANCE.   The current code coverage of Compressed Natural Gas (CNG) motor vehicle fuel dispensing focuses on commercial and private fleet installations. New requirements have been added recognizing the residential application of CNG fuel dispensing consistent with the requirements of NFPA 52, Compressed Natural Gas (CNG) Vehicular Fuel Systems. With appropriate limitations and safeguards, residential CNG appliances are allowed in locations not permitted under the requirements for commercial installations.

Residential CNG fueling appliances must be listed and are limited to a delivery capacity of not more than 5 standard cubic feet per minute of natural gas. When installed outdoors, the appliance must be

413.2.3 continues

413.2.3 continued on a firm, noncombustible base. A new exception has been added to the general requirement requiring an outdoor location. Residential appliances may be located indoors when installed in accordance with the manufacturer's installation instructions and the restrictions of Section 413.4.3: (1) the appliance must be vented to the outdoors and (2) a gas detector must be installed in the room or space containing the appliance. The detector must be located within 6 inches of the highest point of the ceiling and shut down the appliance when activated. The detector alarm may be audible or visual.

The code clarifies that the limitations of Section 413.3.1 for location on property do not apply to residential fueling appliances.

CHANGE TYPE.   Addition

CHANGE SUMMARY.   New requirements have been added for over-pressure protection devices for high-pressure gas systems, which would apply mainly to large commercial and industrial installations. The code is now consistent with requirements of ANSI Z223.1.

2006 CODE: <u>**Section 416 (IFGS) Overpressure Protection Devices**</u>

416.1 General.   <u>Overpressure protection devices shall be provided in accordance with this section to prevent the pressure in the piping system from exceeding the pressure that would cause unsafe operation of any connected and properly adjusted gas utilization equipment.</u>

416.2 Protection Methods.   <u>The requirements of this section shall be considered to be met and a piping system deemed to have over-pressure protection where a service or line pressure regulator plus one other device are installed such that the following occur:</u>

1. <u>Each device limits the pressure to a value that does not exceed the maximum working pressure of the downstream system.</u>
2. <u>The individual failure of either device does not result in over-pressurization of the downstream system.</u>

416 continues

416

Overpressure Protection Devices

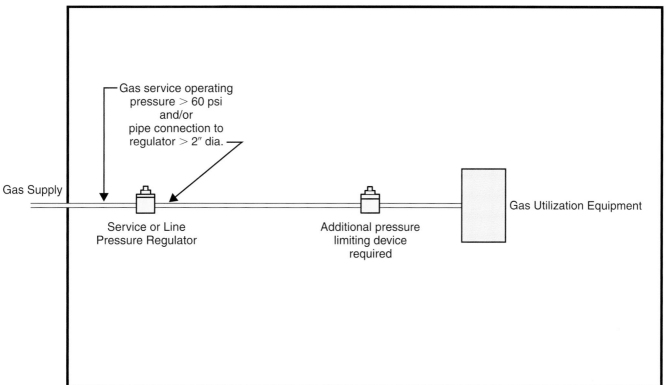

**Overpressure Protection Devices for
High Pressure Industrial Application**

416 continued

416.3 Device Maintenance. The pressure-regulating, -limiting, and -relieving devices shall be properly maintained, and inspection procedures shall be devised or suitable instrumentation installed to detect failures or malfunctions of such devices, and replacements or repairs shall be promptly made.

416.4 Where Required. A pressure-relieving or -limiting device shall not be required where (1) the gas does not contain materials that could seriously interfere with the operation of the service or line pressure regulator; (2) the operating pressure of the gas source is 60 psi (414 kPa) or less; and (3) the service or line pressure regulator has all of the following design features or characteristics:

1. Pipe connections to the service or line regulator do not exceed 2 in. (51 mm) nominal diameter.
2. The regulator is self-contained with no external static or control piping.
3. The regulator has a single port valve with an orifice diameter no greater than that recommended by the manufacturer for the maximum gas pressure at the regulator inlet.
4. The valve seat is made of resilient material designed to withstand abrasion of the gas, impurities in the gas, and cutting by the valve and to resist permanent deformation where it is pressed against the valve port.
5. The regulator is capable, under normal operating conditions, of regulating the downstream pressure within the necessary limits of accuracy and of limiting the discharge pressure under no-flow conditions to not more than 150% of the discharge pressure maintained under flow conditions.

416.5 Devices. Pressure-relieving or pressure-limiting devices shall be one of the following:

1. Spring-loaded relief device.
2. Pilot-loaded back pressure regulator used as a relief valve and designed so that failure of the pilot system or external control piping will cause the regulator relief valve to open.
3. A monitoring regulator installed in series with the service or line pressure regulator.
4. A series regulator installed upstream from the service or line regulator and set to continuously limit the pressure on the inlet of the service or line regulator to the maximum working pressure of the downstream piping system.
5. An automatic shutoff device installed in series with the service or line pressure regulator and set to shut off when the pressure on the downstream piping system reaches the maximum working pressure or some other predetermined pressure less than the maximum working pressure. This device shall be designed so that it will remain closed until manually reset.

6. A liquid seal relief device that can be set to open accurately and consistently at the desired pressure.

The devices shall be installed either as an integral part of the service or line pressure regulator or as separate units. Where separate pressure-relieving or pressure-limiting devices are installed, they shall comply with Sections 416.5.1 through 416.5.6.

416.5.1 Construction and Installation. Pressure-relieving and pressure-limiting devices shall be constructed of materials so that the operation of the device will not be impaired by corrosion of external parts by the atmosphere or of internal parts by the gas. Pressure-relieving and pressure-limiting devices shall be designed and installed so that they can be operated to determine whether the valve is free. The devices shall also be designed and installed so that they can be tested to determine the pressure at which they will operate and examined for leakage when in the closed position.

416.5.2 External Control Piping. External control piping shall be protected from falling objects, excavations, and other causes of damage and shall be designed and installed so that damage to any control piping will not render both the regulator and the overpressure protective device inoperative.

416.5.3 Setting. Each pressure-relieving or pressure-limiting device shall be set so that the pressure does not exceed a safe level beyond the maximum allowable working pressure for the connected piping and appliances.

416.5.4 Unauthorized Operation. Precautions shall be taken to prevent unauthorized operation of any shutoff valve that will make a pressure-relieving valve or pressure-limiting device inoperative. The following are acceptable methods for complying with this provision:

1. The valve shall be locked in the open position. Authorized personnel shall be instructed in the importance of leaving the shutoff valve open and of being present while the shutoff valve is closed so that it can be locked in the open position before leaving the premises.
2. Duplicate relief valves shall be installed, each having adequate capacity to protect the system, and the isolating valves and three-way valves shall be arranged so that only one safety device can be rendered inoperative at a time.

416.5.5 Vents. The discharge stacks, vents, and outlet parts of all pressure-relieving and pressure-limiting devices shall be located so that gas is safely discharged to the outdoors. Discharge stacks and vents shall be designed to prevent the entry of water, insects, and other foreign material that could cause blockage. The discharge stack or vent line shall be at least the same size as the outlet of the pressure-relieving device.

416 continues

416 continued

416.5.6 Size of Fittings, Pipe, and Openings. The fittings, pipe and openings located between the system to be protected and the pressure-relieving device shall be sized to prevent hammering of the valve and to prevent impairment of relief capacity.

CHANGE SIGNIFICANCE. The change brings provisions into the code for overpressure protection in high-pressure gas systems, including those exceeding 60 psi, employed in large industrial and commercial installations. The code was previously silent in this regard and now is consistent with the language found in ANSI Z223.1. Overpressure protection devices are designed to prevent pressure in the gas piping system from exceeding safe limits. Redundancy is provided by requiring two devices on a system, one of which must be a service or line pressure regulator. Each device limits the downstream system to a safe working pressure should the other device fail. The code also requires monitoring and maintenance of the pressure protection devices.

Under certain circumstances, the additional device beyond the service or line pressure regulator is not required. Such a pressure-relieving or -limiting device is not required, provided the gas contains no foreign materials injurious to the service or line pressure regulator, the operating pressure is 60 psi or less, the pipe connection to the service or line regulator does not exceed 2 inches in diameter, and a number of other specific performance specifications are met by the service or line pressure regulator.

The code now contains very specific design and installation requirements for the pressure-limiting or pressure-relieving devices. Precautions are required to prevent unauthorized operation of shut-off valves that may impair the operation of the devices, including an option for locking valves in the open position as satisfying this requirement.

Venting of all pressure-relieving and pressure-limiting devices must safely discharge gas to the outside atmosphere and prevent the entry of insects, water, and foreign material.

CHANGE TYPE. Addition

CHANGE SUMMARY. In this new provision, concealed vents installed through framing members must be protected from fastener penetration by a steel shield plate unless the distance from the face edge of the framing is not less than 1.5 inches.

2006 CODE: <u>**502.7 Protection Against Physical Damage.** In concealed locations, where a vent is installed through holes or notches in studs, joists, rafters or similar members less than 1.5 inches (38 mm) from the nearest edge of the member, the vent shall be protected by shield plates. Shield plates shall be a minimum of $\frac{1}{16}$-inch-thick (1.6 mm) steel, shall cover the area of the vent where the member is notched or bored, and shall extend a minimum of 4 inches (102 mm) above sole plates, below top plates and to each side of a stud, joist, or rafter.</u>

CHANGE SIGNIFICANCE. Protection is now required for concealed flue vents, such as Type B or PVC pipe vents, passing through framing materials and less than 1.5 inches from the face of the framing member. Fuel gas piping and gaseous hydrogen piping subject to anchor penetration similarly require protection in accordance with Sections 404.5 and 704.1.2.3.5. This also brings consistency with the mechani-

502.7 continues

502.7

Protection of Vents Against Physical Damage

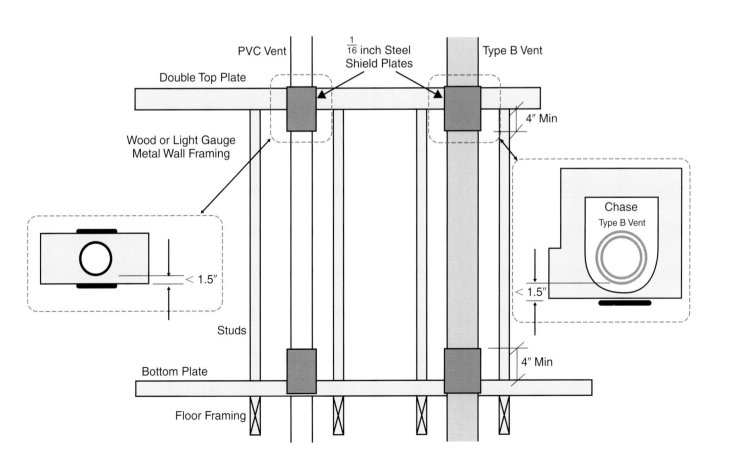

502.7 continued   cal and plumbing piping protection provisions of the IRC, IMC, and IPC. It is appropriate to provide the same level of protection to the integrity of vent systems carrying flue gases and the byproducts of combustion.

In a typical residential drywall application, a common $1\frac{5}{8}$–inch drywall screw attaching $\frac{1}{2}$-inch gypsum board would penetrate the framing member at least $1\frac{1}{8}$ inches plus countersinking depth. Similarly, nails from a pneumatic nailer attaching paneling or trim may penetrate the framing to a depth greater than 1 inch, depending on the length of the nail and depth of the countersink. Placing the vent at least 1.5 inches back from the face of the framing member or protecting it with a steel shield plate is consistent with the other code sections cited and is judged to be a reasonable precaution in light of the above examples.

Unlike Sections 404.5 and 704.1.2.3.5, the protection requirements of Section 502.7 are not limited to installations in wood framing but would apply in light-gauge steel framing or similar materials.

CHANGE TYPE. Modification

CHANGE SUMMARY. The code now prohibits gas piping to bypass a solenoid valve installed as part of the required interlock system of kitchen exhaust hoods serving as appliance vents. An exception to the appliance and exhaust interlock requirement has been added to allow an automatically activated exhaust system.

2006 CODE: 505.1.1 Commercial Cooking Appliances Vented by Exhaust Hoods. Where commercial cooking appliances are vented by means of the Type I or Type II kitchen exhaust hood system that serves such appliances, the exhaust system shall be fan powered and the appliances shall be interlocked with the exhaust hood system to prevent appliance operation when the exhaust hood system is not operating. Where a solenoid valve is installed in the gas piping as part of an interlock system, gas piping shall not be installed to bypass such valve. Dampers shall not be installed in the exhaust system.

505.1.1 continues

505.1.1

Commercial Cooking Appliances Vented by Exhaust Hoods

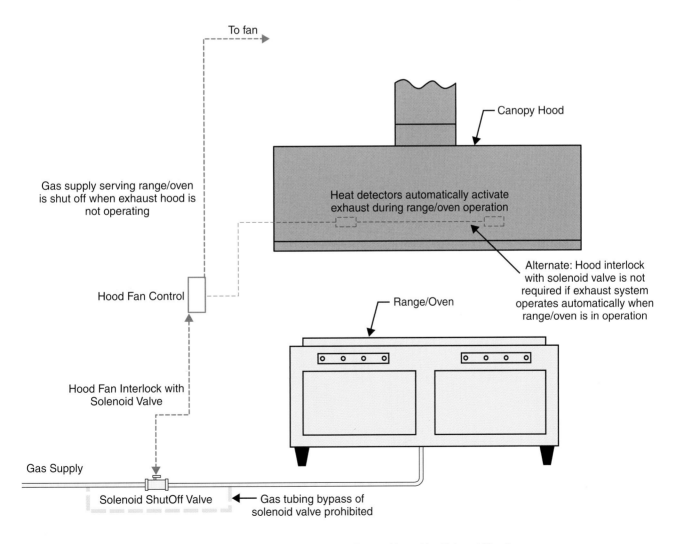

Commercial Cooking Appliances Vented by Exhaust Hoods

505.1.1 continued

Exception: An interlock between the cooking appliance(s) and the exhaust hood system shall not be required where heat sensors or other approved methods automatically activate the exhaust hood system when cooking operations occur.

CHANGE SIGNIFICANCE. In commercial kitchen applications, Type I hoods are required above appliances producing grease-laden vapors and Type II hoods are used to exhaust heat and steam absent of grease and smoke. The requirements and installation methods for kitchen exhaust hoods are found in Chapter 5 of the IMC. Type I and Type II hoods also serve as the required vent for combustion products of gas-fired commercial cooking appliances, in accordance with IFGC Sections 501.2, 503.3.4, and 505.1.1. A gravity vent is not allowed when using a kitchen hood but there must be a fan-powered exhaust system. To prevent the hazardous accumulation of combustion by-products, the exhaust system must be interlocked to the operation of the appliance, such as a range, to insure proper venting whenever the appliance is being used. This is typically accomplished with a solenoid valve on the gas piping serving the appliance. The valve shuts off the gas supply to the appliance if the hood exhaust system is not operating.

The code now specifically prohibits any bypass of the solenoid valve on the gas piping. In some installations, small-diameter gas tubing or piping was installed to bypass the solenoid valve to allow standing pilot burners to operate continuously. Without the bypass, the pilots must be manually lit after each shut-down of the exhaust system. The bypass creates a potentially dangerous situation in the operation of the affected appliance. The small tube or pipe could act as a severely undersized connector that supplies an unknown rate of gas flow to open burners, creating the potential for hazardous appliance operation, delayed ignition, or the buildup of fuel gas.

A new exception provides that the interlock between the exhaust and appliance operation is not required if the exhaust system activates automatically when the appliance is used. The means of activation may be heat sensors in the hood or some other approved means. This provision has been added to specifically address the manual operation of a commercial range or other cooking appliance with standing pilots. An interlock system prevents the continuous operation of the pilot burners, as intended in the design of the appliance, and they must be relit after each shut-down of the exhaust system. Commercial cooking appliances are often designed such that pilots are not easily accessible for frequent relighting, increasing the risk of injury to personnel. The possibility also exists that the equipment will be field retrofitted for ease in relighting, voiding the design certification and creating potential hazards. The code recognizes that the operation of the exhaust is critical in venting flue gases of the cooking appliance and should not rely on the manual activation by kitchen staff. Automatic activation of the hood exhaust system provides an appropriate level of safety while allowing the pilots to burn continuously as intended in the design of the appliance.

CHANGE TYPE.   Modification

CHANGE SUMMARY.   The language prohibiting storage of more than three hydrogen-fueled vehicles in a garage or service station used for hydrogen generation or refueling has been removed. Requirements for maximum floor area, hydrogen flow rate, ventilation, and openings related to indoor locations for hydrogen operations have been clarified. A new provision requires all ignition sources be located below mechanical ventilation outlets.

2006 CODE: 703.1 Hydrogen-Generating and Refueling Operations.   Ventilation shall be required in accordance with Section 703.1.1, 703.1.2, or 703.1.3 in public garages, private garages, repair garages, automotive motor fuel-dispensing facilities and parking garages ~~which~~ that contain hydrogen-generating appliances or refueling systems. ~~Such spaces shall be used for the storage of not more than three hydrogen-fueled passenger motor vehicles and have a floor area not exceeding 850 square feet (79 m²). The maximum rated output capacity of hydrogen-generating appliances shall not exceed 4 standard cubic feet per minute (ft³/min) of hydrogen for each 250 square feet (23.2 m²) of floor area in such spaces. Such equipment and appliances shall not be installed in Group H occupancies except where the occupancy is specifically designed for hydrogen use, or in control areas where open use, handling or dispensing of combustible, flammable or~~

703.1 continues

703.1

Hydrogen-Generating and Refueling Operations

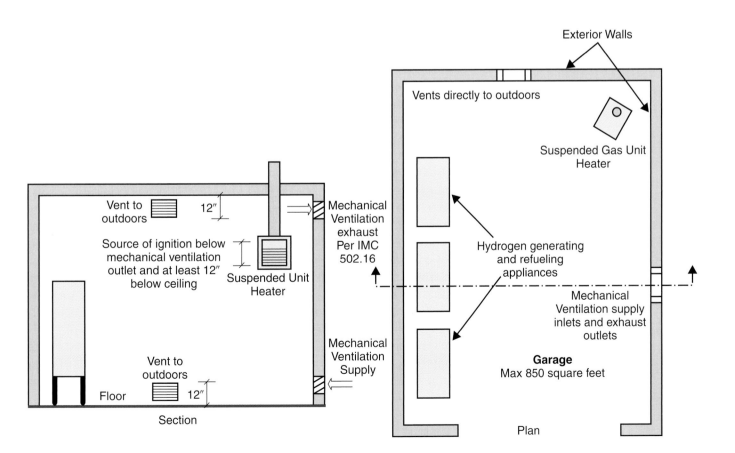

703.1 continued   ~~explosive materials occurs.~~ For the purpose of this section, rooms or spaces that are not part of the living space of a dwelling unit and that communicate directly with a private garage through openings shall be considered to be part of the private garage.

703.1.1 Natural Ventilation. Indoor locations intended for hydrogen-generating or refueling operations shall <u>be limited to a maximum floor area of 850 square feet (79 m²) and shall</u> communicate with the outdoors in accordance with Sections 703.1.1.1 through 703.1.1.2. <u>The maximum rated output capacity of hydrogen generating appliances shall not exceed 4 standard cubic feet per minute (0.00189 m³/s) of hydrogen for each 250 square feet (23.2 m²) of floor area in such spaces.</u> The minimum cross-sectional dimension of air openings shall be 3 inches (76 mm). Where ducts are used, they shall be of the same cross-sectional area as the free area of the openings to which they connect. In such locations, equipment and appliances having an ignition source shall be located such that the source of ignition is not ~~less than~~ <u>within</u> 12 inches (305 mm) ~~below~~ <u>of</u> the ceiling.

703.1.1.1 Two Openings.   Two permanent openings~~,~~ <u>shall be provided within the garage. The upper opening shall be</u> ~~one~~-located entirely within 12 inches (305 mm) of the ceiling of the garage~~,~~<u>. The lower opening shall be</u> ~~and one~~-located entirely within 12 inches (305 mm) of the floor of the garage~~,~~<u>. Both openings</u> shall be provided in the same exterior wall. The openings shall communicate directly with the outdoors~~. Each opening shall directly communicate with the outdoors horizontally,~~ and <u>shall</u> have a minimum free area of ½ square foot per 1000 cubic feet (1 m²/610 m³) of garage volume.

703.1.2 Mechanical Ventilation. Indoor locations intended for hydrogen-generating or refueling operations shall be ventilated in accordance with Section 502.16 of the *International Mechanical Code.* <u>In such locations, equipment and appliances having an ignition source shall be located such that the source of ignition is below the mechanical ventilation outlet(s).</u>

CHANGE SIGNIFICANCE.   The number of hydrogen-fueled vehicles is no longer limited in garages that contain hydrogen-generating appliances or refueling systems. The degree of hazard is more appropriately addressed by limiting the floor area, the output rate of hydrogen-generating appliances in proportion to the area of the space, and proper ventilation. The references to Group H occupancies and control areas have been removed. Section 305.2 still prohibits appliances with an ignition source from being installed in Group H occupancies or in control areas where open use of combustible, flammable, or explosive materials occurs.

Hydrogen is buoyant and disperses readily. Should there be a leak of hydrogen gas, ventilation is necessary to prevent the accumulation of the flammable gases in the space. Though the language has been clarified, the requirements for ventilation opening size and location remain the same.

Section 305.3 addresses the hazards of gasoline vapors and other heavier-than-air flammables in garages by requiring elevation of ignition sources 18 inches above the floor. Similarly, Section 703.1 addresses the hazards of lighter-than-air hydrogen gas and natural gas (often used in the production of hydrogen) that may collect near the ceiling. To prevent the possibility of an explosion, for any appliance with a source of ignition installed in a hydrogen-generating or refueling space, the source of ignition must be located at least 12 inches below the ceiling and located below any mechanical ventilation outlets.

704.1.2.3

Hydrogen Piping System Design and Installation

CHANGE TYPE. Modification

CHANGE SUMMARY. All 300 series stainless steel piping and tubing is now approved for hydrogen gas. Hydrogen gas piping is no longer allowed in concealed locations.

2006 CODE: 704.1.2.3 Piping Design and Construction. Piping ~~systems~~ and tubing materials shall be ~~Type 304, Type 304L or Type 316~~ 300 series stainless steel ~~tubing~~ or materials listed or approved for hydrogen service and the use intended through the full range of ~~pressure and temperature~~ operating conditions to which they will be subjected. Piping systems shall be designed and constructed to provide allowance for expansion, contraction, vibration, settlement and fire exposure.

704.1.2.3.1 Prohibited Locations. Piping shall not be installed in or through a circulating air duct, clothes chute, chimney or gas vent, ventilating duct, dumbwaiter or elevator shaft. <u>Piping shall not be concealed or covered by the surface of any wall, floor or ceiling.</u>

704.1.2.3.2 Interior Piping. <u>Except for through penetrations, piping located inside of buildings shall be installed in exposed locations and provided with ready access for visual inspection.</u>

Hydrogen Piping System Design and Installation Approved Materials

Piping and Tubing Material	Type	Use
Stainless Steel	302	Approved for Hydrogen System
	314	
	314L	
	316	
	316L	
	300 Series	
Black Steel, Plastics, or Other Materials		Only if specifically **listed or approved** for the application and full range of operating conditions
Cast Iron		**Prohibited** from use in hydrogen piping systems

- The code recognizes the use of any of the austenitic stainless steel alloys in the 300 series, which meet the temperature limits of ASME B31.3.
- Hydrogen piping shall not be concealed in any wall, floor, or ceiling.
- Piping located inside of buildings shall be exposed with ready access for visual inspection.
- Underground piping shall be protected from corrosion.
- Piping systems shall be designed for expansion, contraction, vibration, settlement, and fire exposure.

704.1.2.3.3 Underground Piping. <u>Underground piping, including joints and fittings, shall be protected from corrosion and installed in accordance with approved engineered methods.</u>

~~**704.1.2.3.2 Piping in Solid Partitions and Walls.** Concealed piping shall not be located in solid partitions and solid walls, except where installed in a ventilated chase or casing.~~

~~**704.1.2.3.3 Piping in Concealed Locations.** Portions of a piping system installed in concealed locations shall not have unions, tubing fittings, right or left couplings, bushings, compression couplings and swing joints made by combinations of fittings.~~

> ~~**Exceptions:**~~
> ~~1. Tubing joined by brazing.~~
> ~~2. Fittings listed for use in concealed locations.~~

~~**704.1.2.3.6 Piping in Solid Floors.** Piping in solid floors shall be laid in channels in the floor and covered in a manner that will allow access to the piping with a minimum amount of damage to the building. Where such piping is subject to exposure to excessive moisture or corrosive substances, the piping shall be protected in an approved manner. As an alternative to installation in channels, the piping shall be installed in a casing of Schedule 40 steel, wrought iron, PVC or ABS pipe with tightly sealed ends and joints and the casing shall be ventilated to the outdoors. Both ends of such casing shall extend not less than 2 inches (51 mm) beyond the point where the pipe emerges from the floor.~~

CHANGE SIGNIFICANCE. Revised language clarifies that hydrogen piping systems include both piping and tubing materials. Compressed gas industry standards are now reflected in the code to recognize the use of any of the austenitic stainless steel alloys in the 300 series, which meet the temperature limits of ASME B31.3. The restriction to only Types 304, 304L, and 316 failed to recognize the common use of other available series 300 stainless steel materials. The code still allows other piping and tubing materials specifically listed or approved for use in hydrogen systems.

The code now clarifies that the listing or approval of materials other than series 300 stainless steel must be based on the full range of operating conditions to which the system will be subjected. The phrase "pressure and temperature" has been replaced because it did not take into account other operating conditions such as pressure cycling.

Hydrogen gas piping inside buildings is no longer allowed in any concealed location (other than the portion hidden in the through-penetration of a wall, floor, or ceiling assembly) but must be readily exposed and visible. The original code provisions were modeled after natural gas installations with relatively low pressures (less than 5 psig). With the development of hydrogen systems employing much higher pressures, the new requirements will afford a higher level of safety in indoor hydrogen systems.

704.1.2.3 continues

704.1.2.3 continued   A new Section 704.1.2.3.3 has been added to provide for the appropriate protection from corrosion for underground hydrogen piping and requires that such underground systems shall be installed in accordance with approved engineered methods. The requirement is similar to corrosion protection requirements for all fuel gas piping found in Section 404.8.

705

Testing, Inspection, and Purging of Hydrogen Piping Systems

CHANGE TYPE.   Modification

CHANGE SUMMARY.   Section 705 has been extensively rewritten to address specific and detailed requirements for testing, inspection, and purging of hydrogen piping systems. Because of its permissive language, ASME B31.3 is no longer mandated by the code, but the standard is now referenced as one option for the designer to specify for testing and inspection if approved by the code official. Prescriptive and enforceable language is now placed in the text of the code and includes options for hydrostatic and pneumatic pressure leak testing, as well as requirements for purging the piping system following testing.

2006 CODE: **Section 705 (IFGC) Testing, Inspection, and Purging of Hydrogen Piping Systems**

705.2 Inspections.   Inspection shall consist of a visual examination of the entire piping system installation and a pressure test~~., prior to system operation. Engineered~~ Hydrogen piping systems shall be ~~designed using approved engineering methods and the inspection procedures of ASME B31.3, and such inspections shall be verified by the code official.~~ inspected in accordance with this code. Inspection methods such as outlined in ASME B31.3 shall be permitted when specified by the design engineer and approved by the code official. Inspections shall be conducted or verified by the code official prior to system operation.

705 continues

Pneumatic Leak Tests for hydrogen piping systems with maximum working pressures less than 125 psig		
Maximum Working Pressure psig	Test Pressure psig	Testing Gauge Increments psi
	1.5 × max working pressure	
120	180	2
110	165	2
100	150	2
90	135	2
80	120	2
70	105	2
60	90	1
50	75	1
40	60	1
30	45	1
20	30	1
10	15	1
5	7.5	0.1

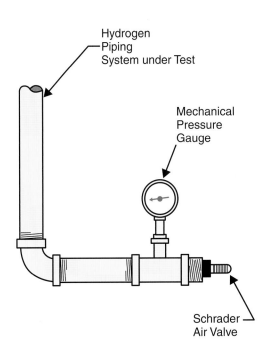

Hydrogen Piping System under Test

Mechanical Pressure Gauge

Schrader Air Valve

705 continued

705.3 Pressure Tests. ~~The test pressure shall be not less than 1½ times the proposed maximum working pressure, but not less than 5 pounds per square inch gauge (psig) (34.5 kPa gauge), irrespective of the design pressure. Where the test pressure exceeds 125 psig (862 kPa gauge), the test pressure shall not exceed a value that produces hoop stress in the piping greater than 50% of the specified minimum yield strength of the pipe.~~ A hydrostatic or pneumatic leak test shall be performed. Testing of ~~engineered~~ hydrogen piping systems shall utilize ~~the~~ testing procedures identified in ASME B31.3 or other approved methods, provided ~~that test duration and gauge accuracy are included in the procedures as~~ the testing is performed in accordance with the minimum provisions specified in Sections 705.3.1 ~~and~~ through ~~705.3.2~~ 705.4.1.

705.3.1 Hydrostatic Leak Tests. The hydrostatic test pressure shall be not less than one-and-one-half times the maximum working pressure, and not less than 100 psig (689.5 kPa gauge).

705.3.2 Pneumatic Leak Tests. The pneumatic test pressure shall be not less than one-and-one-half times the maximum working pressure for systems less than 125 psig (862 kPa gauge), and not less than 5 psig (34.5 kPa gauge) whichever is greater. For working pressures at or above 125 psig (862 kPa gauge), the pneumatic test pressure shall be not less than 110 percent of the maximum working pressure.

705.3.3 Test Limits. Where the test pressure exceeds 125 psig (862 kPa gauge), the test pressure shall not exceed a value that produces hoop stress in the piping greater than 50 percent of the specified minimum yield strength of the pipe.

705.3.4 Test Medium. Deionized water shall be utilized to perform hydrostatic pressure testing and shall be obtained from a potable source. The medium utilized to perform pneumatic pressure testing shall be air, nitrogen, carbon dioxide, or an inert gas. Oxygen shall not be used.

~~705.3.1~~ 705.3.5 Test Duration. The minimum test duration shall be ½ hour. The test duration shall ~~not~~ be not less than ½ hour for each 500 cubic feet (14.2 m³) of pipe volume or fraction thereof. For piping systems having a volume of more than 24,000 cubic feet (680 m³), the duration of the test shall not be required to exceed 24 hours. The test pressure required in Sections 705.3.1 and 705.3.2 shall be maintained for the entire duration of the test.

~~705.3.2~~ 705.3.6 Test Gauges. Gauges used for testing shall be as follows:

1. Tests requiring a pressure of 10 ~~pounds per square inch (psi)~~ psig (68.95 kPa gauge) or less shall utilize a testing gauge having increments of 0.10 psi (0.6895 kPa) or less.

2. Tests requiring a pressure greater than 10 psig (68.98 kPa gauge) but less than or equal to 100 psig (689.5 kPa gauge)

shall utilize a testing gauge having increments of 1 psi (6.895 kPa) or less.

3. Tests requiring a pressure greater than 100 psig (689.5 kPa gauge) shall utilize a testing gauge having increments of 2 psi (13.79 kPa) or less.

Exception: Measuring devices having an equivalent level of accuracy and resolution shall be permitted where ~~approved~~ specified by the design engineer and approved by the code official.

705.3.7 Test Preparation. Pipe joints, including welds, shall be left exposed for examination during the test.

705.3.7.1 Expansion Joints. Expansion joints shall be provided with temporary restraints, if required, for the additional thrust load under test.

705.3.7.2 Equipment Disconnection. Where the piping system is connected to appliances, equipment, or components designed for operating pressures of less than the test pressure, such appliances, equipment, and components shall be isolated from the piping system by disconnecting them and capping the outlet(s).

705.3.7.3 Equipment Isolation. Where the piping system is connected to appliances, equipment or components designed for operating pressures equal to or greater than the test pressure, such appliances, equipment and components shall be isolated from the piping system by closing the individual appliance, equipment, or component shutoff valve(s).

705.4 Detection of Leaks and Defects. The piping system shall withstand the test pressure specified for the test duration specified without showing any evidence of leakage or other defects. Any reduction of test pressures as indicated by pressure gauges shall indicate a leak within the system. Piping systems shall not be approved except where this reduction in pressure is attributed to some other cause.

705.4.1 Corrections. Where leakage or other defects are ~~located~~ identified, the affected portions of the piping system shall be repaired and retested.

705.5 Purging of Gaseous Hydrogen Piping Systems. Purging shall comply with Sections 705.5.1 through 705.5.4.

705.5.1 Removal from Service. Where piping is to be opened for servicing, addition, or modification, the section to be worked on shall be isolated from the supply at the nearest convenient point, and the line pressure vented to the outdoors. The remaining gas in this section of pipe shall be displaced with an inert gas.

705 continues

705 continued

705.5.2 Placing in Operation. <u>Prior to placing the system into operation, the air in the piping system shall be displaced with inert gas. The inert gas flow shall be continued without interruption until the vented gas is free of air. The inert gas shall then be displaced with hydrogen until the vented gas is free of inert gas. The point of discharge shall not be left unattended during purging. After purging, the vent opening shall be closed.</u>

705.5.3 Discharge of Purged Gases. <u>The open end of piping systems being purged shall not discharge into confined spaces or areas where there are sources of ignition except where precautions are taken to perform this operation in a safe manner by ventilation of the space, control of purging rate, and elimination of all hazardous conditions.</u>

705.5.3.1 Vent Pipe Outlets for Purging. <u>Vent pipe outlets for purging shall be located such that the inert gas and fuel gas is released outdoors and not less than 8 feet (2438 mm) above the adjacent ground level. Gases shall be discharged upward or horizontally away from adjacent walls to assist in dispersion. Vent outlets shall be located such that the gas will not be trapped by eaves or other obstructions and shall be at least 5 feet (1524 mm) from building openings and lot lines of properties that can be built upon.</u>

705.5.4 Placing Equipment in Operation. <u>After the piping has been placed in operation, all equipment shall be purged in accordance with Section 707.2 and then placed in operation, as necessary.</u>

707.2 Purging. Purging of gaseous hydrogen systems<u>, other than piping systems purged in accordance with Section 705.5,</u> shall be in accordance with Section 2211.8 of the *International Fire Code* <u>or in accordance with the system manufacturer's instructions.</u>

Chapter 8 Referenced Standards. ASME B31.3 ~~1999~~ <u>2002</u> Process Piping 704.1.2, 705.2, 705.3

CHANGE SIGNIFICANCE. Previously, the code mandated the inspection procedures of ASME B31.3 for hydrogen piping systems. ASME B31.3, Process Piping, is written in permissive language that does not lend itself to effective and uniform enforcement and does not contain all necessary requirements specific to hydrogen systems. Prescriptive and enforceable language applicable to hydrogen is now contained in the text of Section 705. ASME B31.3, updated to the 2002 version in the referenced standards of Chapter 8, is still recognized for inspection methods if specified by the design engineer and approved by the code official, provided all other specific inspection requirements of the code are met.

Proper testing protocol is essential to the installation of gaseous hydrogen piping system installations, not only to prevent hazardous leaks of hydrogen gas when put into operation but also to provide for safe testing pressure limits for the system design and to ensure that no contaminants remain in the system that would be detrimental to the piping materials, equipment components, or utilization of the hydrogen gas. The code recognizes both hydrostatic pressure testing with deionized water and pneumatic pressure testing with air, nitrogen, carbon dioxide, or an inert gas. For systems with working pressures at or above 125 psig, the pneumatic test pressure has been reduced to not less than 110% of the maximum working pressure (compared to the previous requirement of 150%) to improve personnel safety while conducting the test. Pressure limits on stress in the piping relative to yield strength of the pipe are still applicable in test pressures exceeding 125 psig, but the provisions have been relocated to Section 705.3.3 for clarity. Requirements have been added for purging of gaseous hydrogen piping systems similar to the requirements of Section 406 related to fuel gas piping.

Table 403.3

Required Outdoor Ventilation Air

Single-Dwelling Unit

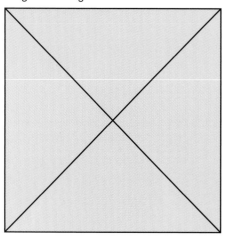

1000 sq. ft. garage
ventilation is not required.

CHANGE TYPE. Modification

CHANGE SUMMARY. Except for garages that serve single family and duplex occupancies, every other garage must now provide exhaust ventilation.

2006 CODE:

TABLE 403.3 Required Outdoor Ventilation Air

Storage		
Repair garages, enclosed parking garages[d]	—	1.5 cfm/ft^2
Warehouses	—	0.05 cfm/ft^2

d. Ventilation systems in enclosed parking garages shall comply with Section 404. ~~A mechanical ventilation system shall not be required in garages having a floor area not exceeding 850 square feet and used for the storage of not more than four vehicles or trucks of 1 ton maximum capacity.~~

(remainder portions of table not shown)

CHANGE SIGNIFICANCE. The change that took place is in the footnote d to Table 403.3. Footnote d in the 2003 *International Mechanical Code* did not require a mechanical ventilation system in a garage having a floor area not exceeding 850 square feet and used for the storage of not more than four vehicles or trucks of 1 ton maximum capacity. The new footnote just refers garage ventilation to Section 404. Section 404.1 was changed to allow intermittent operation of the garage exhaust system when movement by vehicles or occupants is detected but does not have an area limitation for the garage. Section 502.14 in the 2006 *International Mechanical Code* is the only place where mechanical ventilation in a parking garage is required. Exception No. 2 to Section 502.14 exempts mechanical ventilation in a garage that serves one- and two-family dwellings but does not have an area limitation on garages that serve other occupancies and larger residential occupancies, such as triplex and fourplex buildings. Footnote b to "Private dwellings, single and multiple" used for "Garages, common for multiple units" also requires mechanical exhaust for all such garages regardless of the size.

As the original code change proposal was modified during the public hearings, it is not clear if this change in application was completely intended or was intended to clarify that such ventilation is required because footnote b to "Private dwellings, single and multiple" used for "Garages, common for multiple units" in both the 2003 and 2006 *International Mechanical Code* also requires mechanical exhaust for all such garages regardless of the size. As an example, a four-unit apartment building that has an 800 square foot garage will require mechanical ventilation, but a 1000 square foot garage that serves a one- or two-family dwelling will not require mechanical ventilation.

The following information is also relevant and must be noted in this regard:

1. Section 406.1 in the 2006 *International Building Code* does not have any requirements for ventilation in a garage that would be classified as a U occupancy.

2. Section 406.3.12 in the 2006 *International Building Code* will not require mechanical ventilation if the garage has openings to comply as an open parking garage.

3. These requirements and discussions apply to buildings regulated under the *International Building Code* and *International Mechanical Code* because residential buildings under the scope of the *International Residential Code* are regulated under the *International Residential Code* mechanical provisions and are not required to provide ventilation. Such accessory garages must now be limited to a maximum of 3000 square feet based on the definition of accessory structures in the 2006 *International Residential Code*.

[*International Residential Code* Scope: The provisions of the *International Residential Code for One- and Two-Family Dwellings* shall apply to the construction, alteration, movement, enlargement, replacement, repair, equipment, use and occupancy, location, and removal and demolition of detached one- and two-family dwellings and townhouses not more than three stories above-grade in height with a separate means of egress and their accessory structures.]

4-Unit Dwelling

800 sq. ft. garage ventilation required per section 502.14 in the IMC.

Index